Climate Science and AI

Dr. Horen Kuecuekyan

Technics Publications

SEDONA, ARIZONA

115 Linda Vista, Sedona, AZ 86336 USA

https://www.TechnicsPub.com

Edited by Sadie Hoberman
Cover design by Lorena Molinari

First Printing 2025

ISBN, print ed.   9781634626491
ISBN, Kindle ed.   9781634626682
ISBN, PDF ed.   9781634626699

Library of Congress Control Number: 2025931543

Contents

Climate Time Travel: Completely New World

In 1900, Helen and James, esteemed members of London's scientific community, had never anticipated the successful outcome of their experimental cryogenic preservation. Initially driven by academic curiosity rather than a genuine belief in its efficacy, they had volunteered for Dr. Harrison's controversial freezing procedure. Now, in 2050, they found themselves blinking in the unforgiving sunlight, their Victorian sensibilities utterly unprepared for the world that had transpired since their cryostasis.

The first sensation Helen experienced was the oppressive heat. Despite the mild London Spring weather that should have been pleasant, the air felt heavy and suffocating, unlike the crisp April mornings she fondly recalled from her past. James, still adjusting his wire-rimmed spectacles, directed her attention to the peculiar transparent barriers that encircled most of the city's structures. These heat shields, as they would later discover, were ingeniously designed to reflect the intensified solar radiation.

"My dear," James whispered, his scientific mind already racing to comprehend the changes, "I don't think we're in London anymore."

The London they knew had been transformed. The Thames Barrier, a massive engineering marvel that their hosts explained was built in the 1980s, now operated almost constantly to prevent the rising seas from flooding the city. The original barrier had been heightened twice since its construction, standing as a testament to humanity's ongoing battle with the rising waters.

Under the guidance of Dr. Sarah Chen-Martinez, the group traversed the climate-controlled streets of New London, which had undergone a transformative metamorphosis. "Adaptation has been paramount," she spoke, gesturing towards the vertical gardens adorning numerous buildings. "These vertical gardens serve a dual purpose: cooling the city and facilitating local food production. We created these as the transcontinental shipping of food was no longer a viable option in the face of escalating climate extremes."

Helen's grip on Thomas's arm became firm as they passed by a "Climate Refugee Integration Center." The building was abuzz with activity, processing newly arrived refugees from regions rendered uninhabitable by escalating temperatures or flooding. A significant number of these individuals originated from parts of the Middle East and North Africa, where temperatures frequently soared above 50°C (122°F) during the summer months.

James, with his Victorian optimism, asserted, "But surely this cannot be universal? Certainly, some regions must have benefited from the warming?"

In certain regions, particularly in the northern latitudes, agricultural advantages have emerged. Northern Canada and

Russia, where aerial travel was once a common mode of transportation, now contribute significantly to global grain production. Dr. Chen-Martinez maintained a serious tone during the discussion.

However, the transition was not seamless. The soil required several years to develop, and the thawing permafrost released substantial quantities of methane, thereby accelerating the warming process.

As they strolled, Helen meticulously observed the individuals surrounding them. Their attire was notably lighter, crafted from unfamiliar materials that exhibited a remarkable ability to reflect heat and facilitate cooling. Notably, a significant portion of the population sported personal climate control devices—petite, aesthetically pleasing machines that generated a microclimate tailored to the wearer's preferences.

During their initial week, the couple immersed themselves in the profound transformation that had reshaped their world. They delved into the history of the Great Barrier Reef, once a vibrant ecosystem, now largely reduced to a memorial park, with only isolated pockets of surviving coral. They also gained insights into the metamorphosis of Alpine ski resorts, which had transformed into bustling mountain bike parks. Additionally, they explored the disappearance of Maldivian villages, now confined to the pages of digital archives. Furthermore, they gained knowledge about the extensive seawalls that had emerged as protective barriers for coastal cities across the globe.

"What concerns me most," Helen confided in James one evening, "is not the physical changes, which are dramatic indeed. It is the realization that our generation, our time, inadvertently contributed to this, knowingly setting the machinery in motion."

Dr. Sarah Chen-Martinez and the microbiologist Dr. Martha Findler, their hosts, had been gracious enough to provide historical records. The couple dedicated hours to perusing the decades of scientific warnings, the political debates, the international agreements, and the delayed actions that had forged this novel world. They gained insights into the carbon dioxide measurements commencing in the 1950s, which documented the consistent ascent in atmospheric greenhouse gases.

During an evening dinner, Dr. Martha Findler shared a remarkable revelation with them. "We possessed the requisite technological advancements to mitigate the most severe consequences of this crisis by 2020. However, the collective will to implement these changes swiftly eluded us."

James was particularly captivated by the technological transformations that had transpired. The fusion power plants, which now provided a substantial portion of the world's electricity, would have appeared akin to sorcery in his time. The carbon capture facilities, which ingeniously extracted greenhouse gases from the atmosphere, represented engineering feats that surpassed his wildest imagination. Nevertheless, he couldn't help but contemplate the simplicity of reducing coal consumption in his own era.

One month into their new existence, the couple visited the London Climate Museum. The facility documented the planet's transformation over the past 150 years. Helen found herself overcome with emotion as she stood before a display showcasing the animals that had succumbed to extinction during that era. The timeline happened almost precisely at the moment when she and James had entered their suspended animation.

"Behold, James," she remarked, directing his attention to a photograph of a polar bear. "Recall when Captain Ross presented us with his specimens from the Arctic expedition? They are now virtually extinct."

The museum also presented some success stories, including the species saved through intervention, the ecosystems protected and restored, and the communities that had adapted and flourished. However, the overall message was clear: the world had undergone fundamental alterations that could have been averted with earlier and more proactive action.

Their presence in 2050 had not gone unnoticed. Climate historians were particularly eager to interview them, to gain firsthand insights into the mindset of the industrial age that had catalyzed these changes. Helen and James found themselves becoming reluctant experts on their own era, bridging the gap between then and now.

"The truly remarkable aspect," James observed during one such interview, "is that the fundamental scientific principles underlying the greenhouse effect were already comprehended during our time. John Tyndall had elucidated it in 1859, and Svante Arrhenius had even calculated the warming effect of carbon dioxide in 1896. We simply failed to anticipate the magnitude of fossil fuel consumption."

The couple's unique perspective made them invaluable consultants for climate adaptation projects. Their intimate knowledge of Victorian architecture enabled engineers to devise more effective strategies for retrofitting historic buildings to withstand the evolving climate conditions. Helen's expertise in botany proved unexpectedly useful in urban agriculture initiatives, as numerous older plant varieties demonstrated remarkable resilience to heat stress.

Six months into their new life, they attended the annual Climate Adaptation Summit held in the floating conference center in Marseilles. The stark contrast between their former and current environments was evident. Delegates from various countries engaged in discussions on strategies for managing the upcoming decades of climate change, concurrently striving to further reduce emissions.

Helen and James had visited the French Mediterranean numerous times, at least four in Marseille. However, the intense sunlight was almost unbearable. The familiar sight of Marseille's Vieux-Port stretched before them, yet something was profoundly amiss. The ancient limestone buildings remained, but everything else had undergone significant transformation.

Where wooden fishing boats populated the harbor, sleek vessels made of materials they'd never seen before cut through the water. The air buzzed with strange flying devices that darted between buildings like mechanical dragonflies. The Notre-Dame de la Garde still watched over the city from its perch, but now it competed with gleaming towers that seemed to touch the clouds.

"James," Helen whispered, her voice crackling from disuse, "what's happened to this city we loved so much?"

The most recent recollection they held was that frigid night in 1898. They had been walking to the hotel from Tante Simone's New Year's celebration, the mistral wind howling around them, when everything abruptly ceased. The air had crystallized into amber, and consciousness had eluded them.

Almost nine decades later, they stood at the same corner near La Canebière, their attire and countenances unaltered while the world had advanced without them. Helen's voluminous bouffant hairdo and James's impeccably tailored suit starkly contrasted with the unfamiliar surroundings of this contemporary rendition of their cherished city.

A holographic advertisement flickered to life beside them, promoting something called "Neural Interface Enhancement" in brilliant 3D. James instinctively reached for Helen's hand, finding comfort in its familiar warmth. At least they were together in this brave new world.

Passersby observed the vintage ambiance of the establishment, some choosing to capture moments with devices that appeared to materialize from their wrists. The lingering aroma of bouillabaisse emanated from nearby restaurants, serving as a reassuring constant amidst the evolving landscape. However, this familiar scent now blended with unfamiliar spices derived from cuisines that were not yet prevalent during their era.

"We must find answers," James declared, his scientific mind already attempting to comprehend their predicament. However, where does one turn when lost in time? The police station where his friend had served was now a museum dedicated to 20th-century law enforcement. The hospital where Helen was treated for her fractured arm had transformed into a biotechnology research center.

As they continued their walk, the rhythmic clicking of their shoes against the familiar cobblestones echoed through the streets. Each landmark they encountered had undergone subtle transformations. The café where they had shared their first breakfast now dispensed coffee through automated dispensers, while their initial visit hotel had been repurposed into a vertical farm, its walls adorned with cascading vegetation.

A young woman noticed their distress and approached them, speaking in a mix of French and English that had become common in this era. When they explained their situation, her eyes widened with recognition. "You're the Frozen Couple of London! We learned about you in school. Scientists have been studying the atmospheric anomaly that preserved you for decades, but no one expected you to actually wake up!"

As the woman spoke, a small crowd gathered, their devices capturing every moment. Helen and James stood closer together, overwhelmed by the attention, but grateful to learn that their story had not been forgotten. Their narrative had become an integral part of not only London's but also Marseilles' rich tapestry of legends, intertwined with tales of ancient Greek founders and medieval saints.

The authorities were promptly notified, and they were escorted to a facility equipped to facilitate their adjustment to their new French reality. As they sat in the illuminated room, answering an incessant stream of inquiries from scientists and historians, their thoughts invariably returned to the vista of the metamorphosed city.

"At least the sea remains unchanged," Helen remarked softly, observing the eternal Mediterranean through a window that automatically adjusted its tint against the radiant afternoon sun. James gently squeezed her hand, recalling the many times they had witnessed that same sea in their past.

Having lost their world, family, and friends, they found solace in each other's presence. As they gazed upon the city that had grown and transformed, they realized that their bond had endured. They had survived the frozen years intact, and now they held the opportunity to explore this unfamiliar world together, just as they had once discovered the vibrant streets of Marseilles during their many happy visits.

As the sun descended over the harbor, casting hues of orange and pink across the sky, they recognized that certain aspects of their lives remained unaltered by the passage of time. In that realization, they found the fortitude to confront the challenges that lay ahead in this new era. They were confident that they would approach each obstacle with the same determination and success that they had demonstrated in the past.

Back in London, Dr. Chen-Martinez reassured them that the situation was not as dire as it seemed. "Over the past decade, we have made significant strides. Renewable energy now accounts for most of the global energy consumption, and there are indications that atmospheric carbon dioxide levels are beginning to stabilize. The current challenge lies in managing the inevitable changes that are already underway."

Despite the profound changes they experienced, Helen and James found themselves deeply grieving for the world they had known. They mourned the crisp autumn days, the predictable seasons, and the seemingly boundless abundance of nature. However, they also discovered solace in humanity's remarkable resilience and adaptability.

One year after their awakening, Helen and James relocated to one of London's emerging floating communities. They joined a community that had chosen to confront the reality of rising seas rather than resisting them. Their residence, a testament to sustainable architecture, seamlessly adapted to the changing tides. Its walls were adorned with phase-change materials that effectively regulated temperature without the need for active cooling.

"Hello Helen," James said one evening as they observed the sunset through their heat-filtered windows, "I find myself contemplating the perspective of individuals from the year 2050 if they were to have the opportunity to visit our era. Would they harbor resentment towards our lack of knowledge, or would they understand that we were incapable of comprehending the repercussions of our actions?"

Helen considered this as she adjusted their room's climate settings. "Perhaps they would feel as we do now – amazed by how much has changed, saddened by what was lost, but hopeful about human ingenuity and adaptation."

Helen and James, unknowingly time travelers, witnessed the most dramatic transformation in human history. Their presence in 2050 served as a tangible reminder of the rapid pace of global change, emphasizing the significance of learning from the past while formulating a vision for the future.

As they settled into their new lives, they gradually found themselves increasingly at peace with their unique roles. They served as intermediaries between two realms, facilitating the comprehension of the past by future generations while simultaneously utilizing their Victorian-era knowledge to contribute to climate adaptation initiatives. Their narrative became a compelling reminder that the actions of each generation reverberate through time, profoundly shaping the world that their descendants inherit.

"We are all time travelers, in a sense," Helen would frequently proclaim during their public lectures. "We merely traverse time at varying velocities. The pertinent inquiry is, what kind of world do we aspire to arrive in?"

Consequently, as Helen and James embarked on their journey through this altered world, their Victorian sensibilities gradually acclimated to the novel circumstances. Their scientific curiosity facilitated their comprehension of the transformations, while their inherent human capacity for optimism enabled them to embrace the future, irrespective of its divergence from the past they had known.

As Dr. Martha Findler and Dr. Sarah Chen-Martinez had said on their first day, "The world didn't end. It just changed. And we changed with it." For Helen and James, who were accidentally there to witness this big shift, those words became their guiding principle as they helped people understand what had happened and how things had changed.

Introduction

This fictional narrative elucidates the anticipated climate evolution based on our current understanding and estimation of climate change until 2050. We posit that our climate will continue to evolve at a similar pace to the past century, and the prevailing scientific calculations are reasonably accurate. The technical content of the book is dedicated to demonstrating the state-of-the-art methodologies of climate science employed in data collection, analysis, and prediction. While we can only assess the reliability of the underlying scientific principles, we cannot predict human behavior on a global scale. Significant changes in either direction, positive or negative, can have profound and unpredictable consequences.

Global climate change significantly impacts various regions of the Earth, including landmasses, oceans, and the atmosphere. These are global areas, as humans, animals, and microorganisms inhabit a constantly evolving environment, leading to corresponding changes in their lives.

The primary objective of this book is not to provide exhaustive details but rather to offer a comprehensive overview of the vast water bodies on the planet. Subsequently, it will delve into the transformation of storm patterns on the Gulf Coast, utilizing the limited scope to elucidate the changes that occurred over decades, transforming the natural state into a perilous condition for the entire ecosystem.

As humans, animals, various organisms, and microorganisms inhabit every part of the Earth, these changes are interconnected and can propagate throughout the system.

Even seemingly insignificant alterations can have substantial or, unfortunately, in very rare cases, catastrophic consequences.

Earth's topography is shaped by a complex interplay between solid Earth processes that deform the Earth within and surface processes that modify the Earth at the surface. Most of Earth's surface area (currently around 71%) constitutes ocean.

Notes to oceans

Rising temperatures in the Pacific Ocean are causing extensive coral bleaching in the Great Barrier Reef. The marine ecosystem is at risk due to ocean acidification.

Furthermore, the rapid changes in Niño patterns have a significant impact on global weather patterns.

In the Atlantic Ocean, the Gulf Stream circulation weakens and sea levels are rising, posing a substantial threat to coastal communities. Rising sea levels are not only affecting island nations but also causing the erosion and disappearance of small islands.

Great Barrier Reef off Cairns coast.

The frequency of intense hurricanes has also increased, with storms of levels 4 and 5, which were once exceptionally rare events, now occurring more frequently. There is a likelihood of these storms occurring multiple times within a decade.

Concurrently, the Arctic ice melt rate has surpassed even the most pessimistic projections.

Monsoon patterns in the Indian Ocean are becoming more fluid and less distinct, with extended dry and wet seasons. This trend is associated with an increase in the frequency and intensity of extreme weather events.

The Southern Ocean is warming at a faster rate than global averages, particularly in depths between 500 and 2000 meters.

This phenomenon has significant impacts on Antarctic ice shelves and marine ecosystems.

Historically, the Southern Ocean has been a major carbon sink, absorbing substantial amounts of carbon dioxide from the atmosphere. However, current warming trends are reducing the ocean's capacity to absorb CO_2, potentially accelerating global warming.

Simultaneously, the warming waters are shifting the Antarctic Circumpolar Current southward, disrupting deep-water formation and altering global ocean circulation patterns. These changes have cascading effects on the ecosystem, causing temperature fluctuations and, consequently, declining krill populations.

Penguin and whale feeding patterns are also affected, as are phytoplankton productivity levels. Invasive species are invading the warmer waters, further exacerbating the ecological disruptions.

The warming of the Arctic Ocean causes a dramatic reduction in the extent and thickness of summer ice, creating ice-free summers by 2050.

The permafrost thawing is extreme as the warming of the Arctic Ocean is happening two to three times faster than the global average.

The Atlantic Ocean

1850-Present: Gulf Stream Changes and Impact

The Gulf Stream, a key component of the Atlantic Meridional Overturning Circulation (AMOC), has undergone significant alterations since the onset of the Industrial Revolution. This analysis examines historical data, discernible trends, and the implications of these transformations for global climate patterns. The evidence suggests a complex interplay between anthropogenic influences and natural variability, potentially leading to far-reaching consequences for global climate systems.

Historical Context and Early Observations

The historical behavior of the Gulf Stream until 1850, prior to the advent of systematic oceanographic measurements, can be reconstructed through various proxy data sources.

Marine sediment cores provide patterns of circulation spanning thousands of years. Ship logs from merchant and naval vessels offer early documentation of current strength. Temperature records from coastal communities provide indirect evidence of Gulf Stream influence, and historical weather patterns recorded in European chronicles correlate with Gulf Stream behavior.

"From 1850 to 1900, early systematic studies of the Gulf Stream emerged, building upon Benjamin Franklin's initial mapping and temperature measurements. The United States Coast Survey, established in 1807, facilitated more frequent observations. In 1855, Matthew Fontaine Maury significantly advanced these efforts by compiling comprehensive wind and current charts. The Challenger Expedition (1872-1876) expanded our knowledge by incorporating deep-ocean measurements.

True systematic oceanographic monitoring began around 1950 with the introduction of standardized measurement techniques and the establishment of regular sampling stations along the North Atlantic. From 1950 to 1980, continuous monitoring stations were established. Satellite altimetry revolutionized our understanding, revealing deep-water formation processes and highlighting the Gulf Stream's crucial role in global climate.

Consolidated and consistent monitoring conducted after 1980 unveiled substantial alterations within the Florida Current, including altered velocity and the emergence of temperature and salinity anomalies, indicating potential impacts of climate change.

Several key parameters have exhibited significant deviations. Surface water temperatures have risen by 0.8°C to 1.2°C since 1900. While deeper waters have warmed at a slower but measurable pace, seasonal temperature variations have become more pronounced. "Hot spots" of elevated temperatures have emerged along the stream's course.

The Gulf Stream's flow has experienced a 15-20% reduction compared to pre-industrial estimates. This reduction has led to altered meander patterns and shifts in the stream's position. These changes in flow rate have impacted mixing patterns, altering circulation and the transport of water.

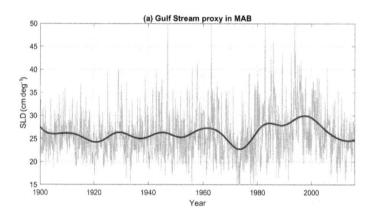

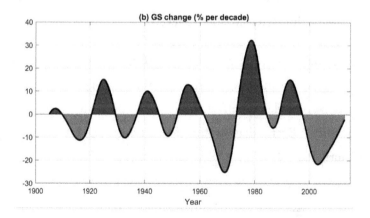

Mid-Atlantic region calculated average sea level changes. The units are percentage changes per decade.

Northern regions have experienced a general freshening trend due to shifts in precipitation and evaporation patterns. Salinity levels along the stream's path exhibit complex patterns, influenced by factors such as increased Arctic ice melt and changes in the evaporation-precipitation balance.

Several prominent features of the Gulf Stream have shifted northward. The meander patterns (twisting course of a flow) of the Gulf Stream have undergone modifications, resulting in the formation of eddies (circular currents transporting water, carbon, salt, and heat from a few to hundreds of kilometers). The width and depth of the Gulf Stream have also been altered.

The alterations in the Gulf Stream have immediate consequences, including modified heat transport to northern latitudes.

Altered precipitation patterns in regions influenced by the Gulf Stream have resulted in changes in storm track positioning and, consequently, impacted coastal sea levels.

These changes have far-reaching global impacts, such as modifying the ocean-atmosphere heat exchange. The absorption of carbon dioxide is affected by alterations to biological productivity. Even deep-water formations exhibit distinct characteristics.

Regional Impacts

Despite global warming, the European climate will exhibit potential cooling. Modified storm tracks and altered precipitation patterns will lead to pronounced seasonal temperature extremes.

North America will experience rising coastal sea levels. These changes will be accompanied by intensified storms and altered precipitation patterns, resulting in increased frequency and severity of extreme weather events.

In the Arctic region, ice melt will accelerate due to modified heat transport, leading to significant changes in the salinity distribution. These alterations will disrupt ecosystem dynamics, causing profound impacts on the region's biodiversity.

Atlantic Meridional Overturning Circulation (AMOC)

The Atlantic Meridional Overturning Circulation (AMOC) is a pivotal component of Earth's climate system, serving as one of the planet's most significant heat distribution mechanisms. This extensive ocean circulation system can be conceptualized as a colossal conveyor belt that transports water across the Atlantic Ocean from the southern regions to the northern hemisphere and back again, exerting a profound influence on global climate patterns and weather systems.

The AMOC comprises two fundamental components: a northward flow of warm, salty surface water originating from the tropics to the northern Atlantic, and a southward flow of colder, deeper water. The surface current, encompassing the Gulf Stream, propels warm water from the Caribbean and South Atlantic toward the North Atlantic. As this water traverses northward, it progressively releases heat into the atmosphere, significantly impacting the climate of Western Europe, thereby maintaining it substantially warmer compared to other regions at comparable latitudes.

Upon reaching the northern Atlantic, particularly in regions near Greenland and the Nordic Seas, the warm, salty water undergoes a significant transformation. The water cools substantially, as cold, salty water is denser than warm water. Consequently, it descends to the ocean depths through a process known as deep water formation. This sinking is a pivotal driver of the entire circulation system, functioning as a pump that propels the global ocean conveyor belt.

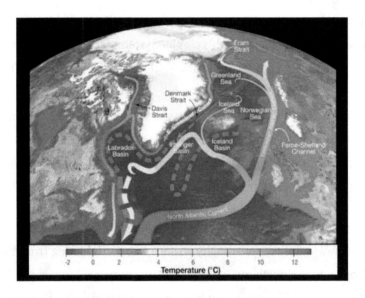

Subsequently, the deep, cold water flows southward at depths of 2-4 kilometers, eventually traversing the Atlantic Ocean. This deep current system eventually surfaces in various regions through upwelling, completing the circulation loop. The entire process transports an immense volume of water—approximately 15-20 million cubic meters per second—making it one of the most extensive water transport systems on Earth.

The Atlantic Meridional Overturning Circulation (AMOC) is of paramount concern to climate scientists due to its potential susceptibility to climate change. As global temperatures increase and Arctic ice melts, the introduction of substantial amounts of freshwater into the North Atlantic could potentially disrupt the circulation by reducing water density and inhibiting the pivotal sinking process. Historical records indicate that significant alterations in the AMOC's strength have been associated with dramatic climate shifts throughout history.

Recent observations suggest that the AMOC has exhibited signs of weakening over the past century, with certain studies proposing that it may be at its weakest point in over a millennium. This weakening could have substantial implications for global climate patterns, potentially influencing precipitation patterns, storm trajectories, and regional temperatures, particularly in areas bordering the North Atlantic.

Research indicates that the Atlantic Meridional Overturning Circulation (AMOC) response to changes in carbon dioxide (CO_2) is fundamentally distinct from the response to freshwater input (FWF). Regardless of the rate of change, an increase in FWF weakens the AMOC, whereas the impact of CO_2 is different. A rapid increase in CO_2 weakens the AMOC, while a very slow (quasi-equilibrium) increase strengthens it.

The Atlantic Meridional Overturning Circulation (AMOC) plays a pivotal role in the global carbon cycle, facilitating the transportation and storage of carbon dioxide in the deep ocean. Consequently, any substantial alterations in its strength have far-reaching implications, not only affecting temperature patterns but also the ocean's capacity to absorb and store carbon dioxide from the atmosphere.

The comprehension and monitoring of the AMOC are paramount for climate science and future climate projections. Scientists employ a diverse range of methodologies to study it, including ocean sensors, satellite observations, and computer models. The RAPID array, a system of moorings strategically positioned across the Atlantic at 26°N, has been providing continuous measurements of the AMOC's strength since 2004, thereby enhancing our understanding of its variability and potential variations.

Ongoing research endeavors to elucidate the dependence and stability of the AMOC on the alterations in CO_2 concentrations under the influence of global warming, an aspect that remains largely unexplored. Furthermore, measurements of the current atmospheric CO_2 state are essential for this purpose.

Short and Long-term Projections

Based on the models we anticipate until 2050, further deceleration of circulation is expected, followed by elevated temperature anomalies, altered precipitation patterns, and accelerated coastal impacts.

For the period beyond 2050 to 2100, the projections are contingent upon emission scenarios. However, it is imperative to monitor any abrupt changes in patterns, cumulative climate effects, and the inability of ecosystems to adapt.

Storm Patterns Hurricanes and Cyclones

Transformation of the Patterns and the Gulf Coast

On August 29, 2021, exactly sixteen years after Hurricane Katrina devastated New Orleans, Hurricane Ida made landfall in Louisiana as a Category 4 storm. With sustained winds of 150 mph and a storm surge that surpassed local levees, Ida served as a stark reminder of the profound impact of climate change on the Gulf Coast. This was not merely another storm but a testament to the new reality that coastal communities must confront in an era of rising temperatures and intensifying weather patterns.

The scientific consensus is no longer theoretical: sea surface temperatures in the Gulf of Mexico have increased by approximately 2°F since 1970, providing more energy to fuel powerful storms. The warmer atmosphere holds more moisture, leading to increased rainfall and flooding during hurricane events.

What were once considered "hundred-year storms" now occur with alarming frequency, compelling communities to reconsider their building codes, evacuation protocols, and other essential measures.

This transformation manifests in diverse forms across the Gulf Coast region.

In Houston, where Hurricane Harvey precipitated over 60 inches of rainfall in 2017, city planners have been compelled to revise flood maps and implement extensive infrastructure projects. The city's narrative exemplifies the challenges faced by urban centers as they adapt to novel weather patterns while concurrently managing rapid growth and development.

Along the Florida Panhandle, where Hurricane Michael's unprecedented intensification in 2018 caught many communities unawares, communities have been compelled to reconstruct with the potential impact of future storms in mind. Building codes have been updated, yet concerns persist regarding the long-term sustainability of coastal development in the face of escalating storm intensity.

In New Orleans, where billions were invested in levee improvements following Hurricane Katrina, residents confront the dual threat of more formidable storms and rising sea levels. The city's experience underscores the limitations of even robust engineering solutions in addressing the mounting challenges posed by climate change.

The human consequences of these alterations extend beyond immediate storm damage. Insurance premiums have surged, making homeownership increasingly unattainable for numerous coastal residents. Some communities face the prospect of managed retreat, the deliberate abandonment of areas deemed excessively vulnerable to rising waters and intensified storms. The economic repercussions reverberate throughout local economies, impacting diverse sectors such as tourism and maritime transportation.

Amidst these challenges, innovative solutions and adaptations emerge. Architects and engineers devise novel approaches to resilient construction. Community organizations establish networks to provide support for vulnerable populations during disasters. Scientists enhance storm prediction and warning systems, while policymakers grapple with the long-term implications of coastal development in the context of climate change.

This narrative transcends an environmental focus; it delves into the human dimension. It involves families contemplating whether to rebuild or relocate after experiencing the loss of their homes. It entails communities grappling with their identity as the very landscape they inhabit undergoes a profound transformation. It encompasses the convergence of science, policy, and daily life as individuals adapt to a new reality shaped by the evolving climate.

The Gulf Coast serves as a harbinger of the challenges that coastal communities worldwide will confront in the decades to come. The manner in which these communities respond, through policy initiatives, technological advancements, and social adaptations, may provide invaluable lessons for the future of human habitation in an era of climate change.

From the verdant sugar cane fields of Louisiana, where the air is thick with the sweet aroma of molasses, to the glittering skyline of Miami, a dazzling metropolis perched on the edge of the turquoise Atlantic, the narrative of intensifying storms is unfolding with increasing urgency. This is not merely a story of shifting sands and eroding shorelines; it is a profound reshaping of life along the entire Gulf Coast.

In the small, close-knit fishing communities of Mississippi, where weathered wooden piers protrude into the churning waters of the Gulf, the threat of hurricanes looms large. Further west, in the industrial heartland of Texas, where towering oil refineries dominate the horizon, the specter of stronger storms casts a long shadow. Understanding this transformation—its causes, its devastating impacts, and the potential pathways to resilience—is no longer an academic pursuit; it is a critical endeavor for anyone seeking to grasp one of the most pressing environmental challenges of our time.

The Evolution of the Hurricanes from 2005 to 2023

The transformation of hurricane patterns along the Gulf Coast can be most effectively elucidated through a meticulous analysis of pivotal storms that have significantly shaped our comprehension of climate change's impact on tropical systems. These instances not only illustrate the escalating intensity of hurricanes but also highlight the intricate interplay of factors that render contemporary hurricanes particularly devastating.

Hurricane Katrina serves as a pivotal benchmark for comprehending the evolution of storm impacts. Upon making landfall as a Category 3 hurricane after attaining Category 5 strength over the Gulf, Katrina's devastating impact was a result of a confluence of its 125 mph winds and catastrophic storm surge that reached heights of up to 28 feet in certain areas. The storm resulted in over 1,800 fatalities and $125 billion in damage, exposing critical vulnerabilities in both infrastructure and emergency response systems.

Katrina's storm surge overwhelmed New Orleans' levee system, which had been engineered to withstand a scenario deemed the most extreme. The failure of these defenses underscored the obsolescence of traditional engineering assumptions in the face of evolving climate conditions. The storm's rapid intensification over unusually warm Gulf waters, a phenomenon that would become increasingly prevalent, caught many by surprise.

Hurricane Harvey in 2023 represented a novel threat: the "rain bomb" hurricane. While Harvey made landfall as a Category 4 storm with 130 mph winds, its most catastrophic impact was attributed to unprecedented rainfall. The storm remained stationary over Houston for several days, accumulating more than 60 inches of precipitation in certain regions. This pattern of slow-moving, rain-intensive storms emerged as a hallmark of climate change's influence on hurricane behavior.

Harvey's rainfall totals surpassed previous U.S. records, with some areas receiving more precipitation in five days than they typically encounter annually. The storm resulted in $125 billion in damages, equaling the cost of Hurricane Katrina. The flooding exposed vulnerabilities created by urban development patterns. Houston's extensive concrete coverage hindered water absorption, while development in flood-prone areas increased the risk of human casualties.

Hurricane Michael in 2024 exhibited a novel pattern of extraordinary rapid intensification. The storm progressed from a tropical depression to a Category 5 hurricane in a mere three days, making landfall with sustained winds of 160 mph. This rapid intensification, driven by anomalously warm Gulf waters, afforded communities insufficient time to prepare effectively.

The impact of Hurricane Michael on Mexico Beach, Florida, where the eye made landfall, was catastrophic. The storm's intensity reduced entire areas to rubble, reducing buildings to concrete foundations. This devastation surpassed conventional building codes and construction practices, necessitating substantial revisions in local regulations.

On the 16th anniversary of Hurricane Katrina, Hurricane Ida (2021 demonstrated the evolving nature of severe weather systems. While New Orleans' enhanced levee system averted catastrophic flooding, Ida's sustained winds of 150 miles per hour inflicted extensive damage to the power grid, resulting in the loss of electricity for over a million customers. Notably, the storm retained its intensity as it traversed inland, causing devastating flooding as far north as New York City. Scientists attribute this pattern of maintained intensity over land to rising global temperatures.

Hurricane Ian in 2022 underscored the escalating economic consequences of severe storms in developed coastal regions. Upon making landfall in Florida with sustained winds exceeding 150 miles per hour and a storm surge of 15 feet, Ian inflicted catastrophic damage on the densely populated Fort Myers area. The storm's insured losses amounted to $50 billion, positioning it as one of the costliest natural disasters in United States history. This calamity precipitated a crisis within Florida's insurance market.

Hurricanes: From Birth to Landfall

The trajectory of an Atlantic hurricane frequently commences thousands of miles from its eventual point of greatest impact. This comprehensive analysis chronicles the evolution of tropical cyclones from their inception as atmospheric disturbances off the western coast of Africa to their transformation into formidable hurricanes within the Gulf of Mexico. Particular emphasis is accorded to the influence of ocean temperatures and currents on this transformative process.

African Easterly Waves (AEWs), low-pressure weather systems, are typically formed over the Ethiopian Highlands and the Sudan region between May and October. These low-pressure weather systems move westward across North Africa into the tropical Atlantic Ocean. On average, AEWs persist for 3-5 days and play a crucial role in shaping rainfall patterns in West Africa.

The primary driving force behind their formation is the intense temperature contrast that develops between the arid Sahara Desert and the more humid Gulf of Guinea coastline. This temperature gradient facilitates the formation of the African Easterly Jet, a concentrated band of winds that forms at approximately 650 mb (millibars) in the atmosphere. Millibars (mb) are units of atmospheric pressure. The standard sea-level pressure is 1013.25 mb. Lower atmospheric pressure corresponds to low-pressure systems, while higher atmospheric pressure corresponds to high-pressure systems. The interaction between this jet and the surrounding environment generates instabilities in the atmospheric flow, primarily through both vertical and horizontal wind shear mechanisms, which ultimately lead to wave formation.

The structure and behavior of a typical African Easterly Wave exhibit distinct characteristics that define its initial potential for further development. These waves typically span vast distances, ranging from 2000 to 4000 kilometers from crest to crest. They move steadily westward across the African continent at speeds averaging 7 to 8 meters per second, establishing a predictable pattern where successive waves pass a given point every 3 to 5 days. The waves reach their maximum amplitude in the middle levels of the atmosphere, typically at approximately 600-700 millibars, where they generate alternating patterns of cyclonic and anticyclonic vorticity that can be observed in weather data.

As these atmospheric waves traverse the African coastline, they undergo a series of pivotal transformations that herald their potential evolution into tropical systems. The most pronounced alteration transpires as the previously arid, continental air mass commences to assimilate maritime moisture from the eastern Atlantic. This process catalyzes the development of shallow convection, wherein warm, moist air ascends and forms clouds. The upward motion augments low-level convergence, drawing more moisture-laden air into the system. Over an extended period, isolated thunderstorms amalgamate into larger, more structured clusters, marking the initial incipient stages of tropical development.

The success or failure of early development is contingent upon a complex interplay of environmental factors. The Saharan Air Layer, a vast expanse of dry, dust-laden air that frequently extends over the Atlantic, can either hinder or facilitate development based on its characteristics and position. The vertical wind shear environment plays a pivotal role, with lower shear generally favoring development by enabling the system to maintain a vertical structure. Sea surface temperatures must be sufficiently warm to provide the requisite energy for development, while mid-level moisture content influences the efficiency of convective processes. Upper-level divergence patterns are essential for ventilating the system and maintaining the vertical circulation necessary for intensification.

The transformation of an African easterly wave into a substantial hurricane is a multifaceted process involving intricate interactions between atmospheric and oceanic processes. This comprehension is fundamental to various aspects of meteorological science and public safety. Firstly, it enables meteorologists to enhance forecast accuracy by more accurately predicting the likelihood of development and intensification. Secondly, it provides a framework for assessing the impacts of climate change on tropical cyclone formation and development. Thirdly, it enhances the preparedness and response capabilities of communities and emergency management organizations to hurricane threats. Finally, it advances our scientific understanding of these intricate atmospheric phenomena.

The anticipated influence of climate change on ocean temperatures and water flow patterns is expected to substantially modify this development process over the coming decades. These alterations are likely to impact several aspects of hurricane development and behavior. Storm formation regions may shift as favorable temperature patterns evolve. Development rates could accelerate in response to warmer waters. Maximum intensities may increase as more energy becomes available to developing systems. Track characteristics could alter as steering currents adjust to changing climate patterns. Furthermore, the duration and timing of the traditional hurricane season may extend as favorable conditions persist for extended periods throughout the year.

This understanding continues to evolve as new observations and research provide deeper insights into these formidable atmospheric phenomena. The ongoing refinement of our knowledge serves both the scientific community and the general public, contributing to enhanced safety and resilience in coastal regions vulnerable to these storms.

The atmospheric conditions over the North Atlantic region are significantly influenced by the Atlantic Meridional Overturning Circulation (AMOC). There is considerable concern regarding the future stability of AMOC. While surface freshwater forcing (FWF) has been extensively researched, it has not been studied in relation to the equilibrium shifts associated with changing and, unfortunately, increasing CO_2 levels. A new approach utilizes a model to explore the stability of atmospheric conditions in CO_2 concentrations between 180 and 560 ppm.

There is growing concern that the Atlantic Meridional Overturning Circulation (AMOC) could weaken or even cease to function due to global warming. Such an event would have substantial implications for the climate of Europe. The potential for irreversible change is a cause for significant concern.

The evolution of the Atlantic Meridional Overturning Circulation (AMOC) is pivotal in comprehending the impact of climate change. The surface ocean freshwater balance is significantly influenced by variations in land ice volume.

Ongoing research suggests that AMOC exhibits distinct responses to alterations in carbon dioxide (CO_2) concentrations compared to freshwater (FWF) forcing. Regardless of the change rate, an increase in FWF leads to a reduction in AMOC, whereas an increase in CO_2 results in a different response pattern: gradual increases enhance AMOC, while rapid increases diminish it.

The Potsdam Institute for Climate Impact Research (PKI) is globally recognized for its significant contributions to climate research. For further reading, a comprehensive scientific account of AMOC is available in the PKI paper.[1]

---

[1] Generalized stability landscape of the Atlantic meridonial overturning circulation Tropical CyclonesMatteo Willeit and Andrey Ganopolski, 2024.

Tropical Cyclones

Tropical cyclones, characterized by intense circular storms forming over warm tropical and subtropical waters, require a series of specific conditions to align for their formation. The ocean water temperature must be at least 26.5°C (80°F) to a depth of approximately 50 meters, as this warm water serves as the primary energy source for the developing storm. The atmosphere must contain high humidity levels in the lower to middle layers, facilitating cloud formation and precipitation. Furthermore, favorable wind conditions are essential, with light winds that maintain relatively consistent speeds and directions with height, known as low wind shear. The Coriolis effect, generated by Earth's rotation, provides the necessary force to initiate and sustain the storm's rotation. Finally, some form of pre-existing atmospheric disturbance or instability is required to trigger the initial development of the tropical cyclone.

The structure of a tropical cyclone is highly organized and distinctive. At its center lies the eye, a remarkably calm region typically spanning 20-40 kilometers in diameter, with light winds and clear skies. Surrounding the eye is the eyewall, which contains the storm's most intense features, including the strongest winds and heaviest rainfall. Spiral rain bands extend outward from the eyewall, rotate around the center, and produce periods of heavy precipitation interspersed with calmer conditions. At the storm's top, high-level winds spread away from the center in a pattern called outflow, which helps ventilate the system and maintain its intensity.

Different regions worldwide employ diverse terminology and classification systems for these cyclonic phenomena. In the North Atlantic and Eastern Pacific, they are referred to as hurricanes, while in the Western Pacific, they are known as typhoons. In the Indian Ocean, they are simply designated as cyclones. The Saffir-Simpson scale, ranging from Categories 1 to 5, is employed to categorize cyclonic intensity based on sustained wind speeds, predominantly in the Atlantic and Eastern Pacific basins.

The impacts of tropical cyclones can be catastrophic and extensive. A storm surge, often the most lethal aspect, entails a surge in sea level that can submerge coastal areas by several meters. High winds, exceeding 250 kilometers per hour in severe systems, can demolish structures, infrastructure, and vegetation. Heavy rainfall frequently leads to extensive inland flooding and landslides, even in regions distant from the coast. The intense wave action generated by these storms poses significant risks to maritime transportation and can severely damage coastal structures.

The life cycle of a tropical cyclone is a well-defined progression. It commences as a tropical disturbance, characterized by heightened thunderstorm activity and subtle indications of rotation. Under favorable atmospheric conditions, it may evolve into a tropical depression, with wind speeds below 63 kilometers per hour. Further intensification propels it to tropical storm status, characterized by wind speeds between 63 and 118 kilometers per hour. At this juncture, the storm is officially named. If it continues to strengthen, it ascends to the category of a full tropical cyclone, hurricane, or typhoon, depending on its location (hurricane in the Atlantic and Pacific Oceans, typhoon in the Northwest Pacific Ocean). Eventually, the storm dissipates, typically when it traverses land or encounters cooler waters.

Modern monitoring and forecasting of tropical cyclones employ a comprehensive suite of sophisticated tools and techniques. Meteorologists rely heavily on satellite imagery to continuously monitor storm development and movement. Aircraft reconnaissance flights provide crucial data by directly sampling conditions within the storm. Networks of weather buoys and radar systems offer additional data regarding wind speeds, pressure, and precipitation. Advanced computer modeling facilitates the prediction of storm tracks and intensity, enabling regular position and intensity forecasts that are paramount for public safety.

In regions prone to tropical cyclones, safety and preparedness are fundamental aspects of residing there. Communities rely on early warning systems to promptly notify residents of approaching storms. Comprehensive evacuation plans facilitate the relocation of individuals from hazardous areas before the impact of the storm. Residents are encouraged to maintain emergency supplies, including food, water, and medical necessities. Building codes in hurricane-prone areas specify construction standards to enhance resilience against high winds and flooding. Storm shutters and other protective measures minimize structural damage.

The interplay between tropical cyclones and climate change is a subject of ongoing scientific research. Scientists have elucidated the potential influences of rising ocean temperatures on various aspects of tropical cyclones. These encompass the possibility of enhanced storm intensity due to increased energy availability from warmer waters, higher rainfall amounts attributable to elevated atmospheric moisture, alterations in the geographic distribution of storm formation and tracking, extended storm seasons, and more severe storm surge impacts due to rising sea levels. These alterations could have significant implications for coastal communities and emergency management planning in the future.

Hurricanes form in the Atlantic Ocean and the Eastern Pacific between June 1st and November 30th, with peak activity typically occurring from mid-August to late October. The Atlantic hurricane season particularly impacts the East Coast and Gulf Coast of the United States, the Caribbean islands, and parts of Mexico and Central America.

These storms originate over warm waters near the equator, frequently off the coast of Africa, and traverse westward toward the Caribbean and North American continent. The Gulf of Mexico and the Caribbean Sea are particularly conducive environments for hurricane development or intensification due to their consistently warm waters.

The National Hurricane Center (NHC) in Miami, Florida, monitors and forecasts these systems for North America. They employ the Saffir-Simpson Hurricane Wind Scale to categorize storms, ranging from Category 1 to Category 5. A Category 1 hurricane possesses sustained winds of 74-95 mph, while a Category 5 hurricane sustains winds exceeding 157 mph. Each progressive category signifies a substantial escalation in the potential for property damage and flooding.

For North America, hurricanes pose distinct threats. Storm surge poses a particularly perilous risk along the Gulf Coast, where the shallow continental shelf and low-lying coastal areas can lead to extensive flooding. The East Coast also faces comparable risks, particularly in regions such as the Outer Banks of North Carolina and the low-lying areas surrounding the Chesapeake Bay.

Rainfall flooding poses a significant threat even in inland regions. For instance, hurricanes that make landfall along the Gulf Coast can track northward, dumping substantial amounts of rain across multiple states. Hurricane Harvey in 2017 exemplified this phenomenon by stalling over Texas and causing catastrophic flooding in Houston and surrounding areas.

Category	m/s	knots	mph	km/h
5	≥ 70	≥ 137	≥ 157	≥ 252
4	58–70	113–136	130–156	209–251
3	50–58	96–112	111–129	178–208
2	43–49	83–95	96–110	154–177
1	33–42	64–82	74–95	119–153
TS	18–32	34–63	39–73	63–118
TD	≤ 17	≤ 33	≤ 38	≤ 62

Saffir-Simpson scale

The United States has established comprehensive hurricane preparedness systems. The Federal Emergency Management Agency (FEMA) collaborates with state and local authorities to manage evacuations and emergency response. The National Weather Service issues watches and warnings, while local emergency management offices coordinate community-level responses. Building codes in hurricane-prone regions mandate specific construction standards, such as impact-resistant windows or elevated foundations in coastal areas.

The Gulf Coast and Atlantic seaboard regions are particularly susceptible to hurricane impacts due to their distinctive geographical and infrastructural features. The Gulf Coast states are particularly vulnerable due to low-lying terrain, subsiding land, and extensive coastal development. New Orleans, situated largely below sea level, faces challenges with aging levee systems and the loss of natural storm protection due to wetland erosion. Houston's sprawling development and impermeable surfaces exacerbate flood risk during heavy rainfall events. Miami's vulnerability stems from its location on porous limestone bedrock, which allows storm surges to penetrate inland through groundwater systems.

Along the Atlantic coast, barrier islands from Florida to North Carolina are subjected to persistent erosion and shifting shorelines, while dense coastal development increases the risk of infrastructure damage. Cities such as Charleston and Norfolk frequently experience "sunny day flooding" even without hurricanes, primarily due to rising sea levels and subsidence. Consequently, these cities are highly vulnerable during hurricane strikes. Many older East Coast cities also have aging stormwater systems that are inadequate to handle the intensity of modern hurricanes. However, they have developed sophisticated evacuation routes and emergency response protocols based on centuries of experience with storms.

Climate change considerations are particularly pertinent to North American hurricanes. Rising sea levels along both the Atlantic and Gulf coasts exacerbate storm surge risks, while warmer waters in the Gulf of Mexico and Atlantic Ocean may contribute to more intense storms. This has necessitated increased focus on coastal resilience planning and infrastructure adaptation in vulnerable regions.

The Arctic

The Arctic tundra, a crucial carbon sink, has undergone significant transformations due to climate change. According to the 2024 Arctic Report of the National Oceanic and Atmospheric Administration (NOAA), it has transitioned into a net source of carbon dioxide emissions.

The accelerated warming of the Arctic is evidence of this alarming trend, which is occurring four times faster than the global average. The ecological consequences are profound, as thawing permafrost exposes ancient organic matter. Consequently, more frequent and intense wildfires in permafrost regions exacerbate carbon emissions, contributing to the intensification of global warming.

Iceberg with hole—Greenland

Approach to Deal with Significant Changes

Modeling challenges include uncertainty in quantification. Scenario development will have gaps because, outside of the Gulf Stream, large events will also alter the modeled patterns.

Enhanced monitoring systems are also necessary, as well as coordination of changes in the Gulf Stream in independent regions. Large wildfires anywhere outside the monitored ocean can significantly alter the model to the point of invalidation.

Missing or changes in emission reduction targets must be continuously incorporated into model creation and updates.

The Gulf Stream's response to climate change is one of the most significant oceanic transformations observed in the contemporary era. The documented alterations in its behavior have far-reaching implications for global climate patterns, marine ecosystems, and human societies. Continued monitoring and research are indispensable for comprehending and adapting to these changes.

The intricate nature of the system and its multifaceted feedback mechanisms render precise predictions arduous. Nevertheless, the discernible trends unequivocally underscore the imperative for sustained attention and proactive measures to address both mitigation and adaptation strategies. The forthcoming decades will be pivotal in determining the long-term stability of this indispensable ocean current system.

Permafrost

Permafrost, the permanently frozen ground that has remained below 0°C for at least two consecutive years, is experiencing substantial alterations globally due to climate change, with particularly pronounced impacts in Arctic and subarctic regions. The thawing of permafrost constitutes one of the most worrisome feedback loops within our climate system.

In regions such as Siberia, Alaska, northern Canada, and certain parts of Scandinavia, permafrost has remained stable for millennia, sometimes extending hundreds of meters beneath the surface. However, as global temperatures increase, these frozen soils are experiencing an unprecedented rate of thawing. This thawing is most evident in the form of thermokarst, where the ground literally collapses as ice within the soil melts, resulting in depressions and creating distinctive landforms and lakes in the landscape.

The primary driver of permafrost thaw is the amplification of global warming in Arctic regions, which are warming at approximately twice the global average rate. This accelerated warming is partly attributable to the albedo effect, wherein the loss of reflective snow and ice surfaces leads to increased absorption of solar radiation by darker land and water surfaces, thereby perpetuating a self-reinforcing cycle of warming.

Permafrost pattern

When permafrost thaws, it releases previously frozen organic matter that has been sequestered for millennia. Consequently, microorganisms initiate the decomposition of this organic material, liberating greenhouse gases, particularly methane and carbon dioxide, into the atmosphere. Scientists estimate that permafrost soils contain approximately twice as much carbon as is currently present in Earth's atmosphere, underscoring the significance of this feedback loop in future climate projections.

The consequences of permafrost thaw extend far beyond the release of greenhouse gases. In numerous Arctic communities, structures and infrastructure built upon previously stable frozen ground are now experiencing structural challenges as their foundations become unstable. Traditional ways of life for indigenous peoples are being disrupted as hunting grounds shift and traditional food storage methods in natural freezers become unreliable.

The thawing also has substantial ecological impacts. As permafrost melts, entire ecosystems are undergoing transformations. Arctic lakes can suddenly drain when the frozen walls containing them thaw. Conversely, new wetlands can form in areas where ground subsidence creates depressions. These changes affect local wildlife, plant communities, and the carbon cycle of these regions.

Coastal erosion is a significant consequence of permafrost thaw. In regions such as the Alaskan and Siberian coasts, the loss of frozen ground that once stabilized the coastline is leading to accelerated erosion, compelling some coastal communities to contemplate relocation. This erosion also contributes additional organic matter to ocean systems, potentially impacting marine ecosystems and biochemical cycles.

The scientific community has observed these changes accelerating in recent decades, with some regions experiencing permafrost temperatures rising by up to 2°C since the 1980s. The warming is particularly pronounced in areas with high ice content in the soil, as the melting of this ice leads to more pronounced physical alterations in the landscape.

Another concerning aspect is the potential release of ancient microorganisms and viruses that have been preserved in the permafrost. While most of these are likely to pose no threat to humans, the scientific community continues to investigate this phenomenon to better comprehend potential risks.

The thawing of permafrost represents a critical juncture in Earth's climate system, and its effects are anticipated to intensify in the forthcoming decades. This underscores the imperative for global action to mitigate greenhouse gas emissions and decelerate the rate of warming, particularly in Arctic regions where the impacts of climate change are most pronounced and immediate.

In addition to these structural changes, we must also recognize that these modifications alter the natural habitat in such a way that microorganisms can now undergo mutations and develop dangerous formations. For any organism, and particularly for microorganisms, temperature, humidity, and density fluctuations in any direction can lead to uncontrolled mutations.

Common Patterns and Trends

Modern hurricanes have fundamentally transformed our understanding of coastal threats and disaster response. Along the Gulf Coast, warmer ocean temperatures have created conditions where storms can intensify dramatically within hours, leaving communities with limited time to prepare. This rapid development pattern represents a significant departure from historical norms when storms typically intensified more gradually.

The reach of hurricane impacts has expanded well beyond traditional coastal zones. Today's storms maintain their destructive power as they move inland, bringing unprecedented rainfall volumes to communities that once considered themselves safely removed from hurricane threats. This expanded impact zone has exposed critical weaknesses in our infrastructure systems. Power grids designed for less severe weather patterns now face repeated failures, while flood control systems struggle to manage the massive water volumes these storms deliver. Traditional engineering standards, once considered robust, increasingly prove inadequate against these intensified threats.

The economic aftermath of modern hurricanes reverberates through communities for years after the storms have passed. Insurance markets are destabilized as they struggle to adjust to new risk patterns, leading to dramatic shifts in coverage availability and costs. Property values fluctuate unpredictably, creating uncertainty in real estate markets that can persist long after physical recovery. Regional economies often take years to fully recover as businesses struggle with reconstruction costs and population shifts in hard-hit areas.

Communities now grapple with complex decisions about their future. The increasing cost of protective infrastructure strains municipal budgets, while the diminishing effectiveness of conventional solutions raises questions about long-term sustainability. Local governments must weigh the substantial costs of rebuilding against the possibility of managed retreat from the most vulnerable areas, all while considering the social and economic implications of their choices.

These evolving patterns demonstrate how climate change has transformed hurricane impact from a primarily coastal concern into a broader regional threat. Each storm provides new lessons that continue to shape emergency management strategies, urban planning decisions, and climate adaptation approaches, not only within the United States but globally.

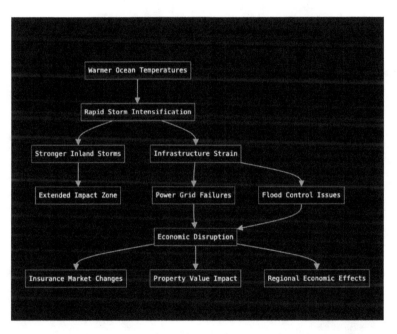

This schematic view shows the significant interdependencies.

CHAPTER ONE

Climate Science AI: Basic Concepts

Analyzing environmental dependencies on climate patterns in any region necessitates the identification of recurring trends. Machine learning and AI) can substantially aid this endeavor, as these technologies are developed based on algorithms designed to discern patterns. Both typical climate events and those influenced by climate change exhibit sequential patterns. These patterns can be analyzed at a high level as time series events.

There are substantial complexities associated with temperature, precipitation, snow coverage, and vegetation.

Conventional Scientific Methods and Approaches

Scientific inquiry commences as a systematic and rigorous methodology for comprehending natural phenomena through observation, hypothesis formulation, experimentation, and analysis. This approach, developed over centuries, provides a framework for generating reliable knowledge about the world.

The scientific process commences with the observation of phenomena and the formulation of precise inquiries. These observations may originate from systematic data collection or from the recognition of anomalous patterns or events in nature.

The subsequent step involves utilizing these observations to formulate testable hypotheses—proposed explanations that can be evaluated through experimentation. A well-constructed hypothesis must be falsifiable, implying its potential for disproven through evidence. Additionally, it should be specific enough to guide experimental design. The hypothesis should establish evidential relationships between variables, such as "Higher water temperatures exceeding X degrees Celsius induce the rate of coral bleaching."

The identification of parameters and variables is a crucial and challenging task, as they must have an impact on the outcome. These are the independent variables that measure the changes in the dependent variables. Experiments designed using these variables must be meticulously crafted to ensure reproducibility, allowing other scientists in the field to replicate the results. Necessary documentation must include detailed methods, materials, and the required conditions for the experiments.

The establishment of control groups is frequently indispensable in most experimental endeavors. The appropriate sample sizes and selection methodologies determine the statistical validity and minimize bias. Control groups are not subjected to the experimental treatment and serve as a baseline for comparison.

Statistical analysis techniques such as confidence intervals, regression analysis, and analysis of variance (ANOVA) assist in ascertaining whether observed differences are statistically significant or attributable to chance.

Utilized instruments must undergo calibration, and the employed techniques must be standardized. Every procedure must be meticulously documented, and observations must be recorded and stored.

When multiple studies consistently support a hypothesis, it may evolve into a theory, a comprehensive explanation supported by substantial evidence. Theories undergo continuous refinement as new evidence emerges. For instance, the theory of plate tectonics developed gradually through accumulated evidence from geology, paleontology, and geophysics.

Scientific theories must elucidate observed phenomena and generate testable predictions. A comprehensive body of evidence must be presented, and it must be receptive to revision in light of the accumulation of new evidence.

Researchers must maintain objectivity and adhere to the evidence, even when it challenges their assumptions. All methods, data, and analyses must be meticulously documented and readily accessible.

The scientific community upholds quality through peer review and evaluation by other experts in the field. This process involves submitting manuscripts to scientific journals for independent review and feedback. This system facilitates the identification of methodological errors, statistical inaccuracies, and alternative interpretations before research is published. The refined hypothesis can then be disseminated.

Limitations and Rules

Scientific methodologies may encounter inherent limitations, such as measurement constraints. Certain phenomena are challenging to measure precisely or directly observe. Furthermore, certain experiments may be ethically prohibitive.

Many natural systems exhibit multiple interacting variables that are difficult to isolate and control.

Researchers must maintain objectivity and adhere to the evidence, even when it deviates from their preconceived notions. All methods, data, and analyses should be meticulously documented and accessible. Results should be verifiable by independent researchers.

Research must adhere to established ethical guidelines, particularly when involving human subjects or animals. Through these systematic approaches, conventional scientific methodologies have yielded reliable knowledge across diverse disciplines, including physics, chemistry, biology, and earth sciences. This framework continues to advance our comprehension of natural phenomena while upholding stringent standards for evidence and analysis.

Environmental Science and AI/ML

Environmental disciplines have extensively explored the application of Machine Learning (ML) and Artificial Intelligence (AI) in identifying opportunities that lie within their domain. The tasks have experienced exponential growth since the advent of deep learning algorithms, which have demonstrated their remarkable potential.

Machine Learning (ML) and AI present a potent synergy that holds the promise of revolutionizing our comprehension and management of the environment. By leveraging the computational capabilities of AI, researchers are gaining unprecedented insights into intricate environmental systems, expediting scientific advancements, and developing innovative solutions to pressing ecological challenges.

> *The most significant impacts of ML and AI lie in their ability to analyze vast and diverse datasets, which are crucial for environmental science due to the enormous volumes of data collected from various sources.*

These include satellite imagery, sensor networks, climate models, and biological observations. Conventional, traditional methods often face challenges in effectively processing and interpreting this data deluge. AI algorithms excel at identifying patterns, anomalies, and hidden relationships within these complex datasets.

Machine learning models can analyze satellite imagery to monitor deforestation rates, track wildlife populations, and assess the health of coral reefs with greater accuracy and speed than manual methods. This enhanced data analysis enables scientists to gain a more comprehensive and nuanced understanding of environmental processes, ranging from local to global scales.

AI is profoundly transforming environmental modeling and prediction. Climate models, pivotal for comprehending and forecasting future climate change, are progressively advancing through the integration of AI methodologies. Machine learning algorithms augment the accuracy and efficiency of these models by elucidating non-linear correlations between variables, optimizing parameter estimation, and downscaling global climate projections to regional and local scales. This facilitates more precise forecasts of extreme weather events, such as floods, droughts, and heatwaves, thereby facilitating enhanced preparedness and mitigation measures.

In addition to data analysis and modeling, AI plays a pivotal role in environmental monitoring and conservation. AI-powered systems can be deployed to monitor air and water quality in real-time, detect and respond to environmental disasters, and optimize resource management. For instance, smart sensors equipped with AI algorithms can monitor water quality in rivers and lakes, promptly alerting authorities to potential pollution events. AI-driven drones can be utilized to survey wildlife populations, track endangered species, and assess the impact of human activities on ecosystems. These applications not only enhance environmental monitoring, but also enable more effective conservation efforts by providing essential data for informed decision-making and intervention.

Integrating AI into environmental science offers substantial benefits while simultaneously presenting challenges and considerations. The potential for bias in AI algorithms is a primary concern arising from biases in the training data or the algorithm's design. If these biases are not addressed, they can lead to inaccurate or misleading results, perpetuating existing environmental injustices and hindering effective conservation efforts.

Therefore, ensuring the fairness, transparency, and accountability of AI systems is paramount for their responsible and ethical utilization in environmental science.

Another significant challenge lies in the computational resources required to train and execute complex AI models. These models often necessitate substantial computing power, which can have environmental implications due to the energy consumption associated with data centers. Researchers are actively exploring strategies to minimize the environmental footprint of AI, including developing more energy-efficient algorithms and utilizing renewable energy sources for data centers.

Despite these challenges, the potential benefits of AI in environmental science are profound. By harnessing the capabilities of AI, researchers can gain a deeper understanding of our planet, develop more effective solutions to environmental challenges, and ultimately contribute to a more sustainable future for all. Continued research and development in this domain are essential to fully realize the transformative potential of AI in addressing the pressing environmental challenges of the 21st century.

Machine learning algorithms can also be utilized to enhance nature conservation efforts. These algorithms can assist in habitat mapping, species distribution modeling, and forest health monitoring. Furthermore, machine learning models can support decision-making processes for establishing no-take zones and implementing protective measures to preserve local biodiversity.

In conclusion, machine learning and artificial intelligence are poised to revolutionize environmental science. By enabling more efficient data analysis, improving predictive models, and facilitating innovative monitoring and conservation strategies, artificial intelligence provides powerful tools for addressing the complex challenges facing our planet. While challenges such as bias and energy consumption must be carefully considered, the potential benefits of artificial intelligence in environmental science are substantial and warrant continued investment and research. By embracing this technological revolution, we can better understand our planet and work towards a more sustainable future.

Challenges in Current Climate Prediction

The Earth's climate system is a highly intricate and multifaceted network comprising various interconnected components, including the atmosphere, oceans, land, ice, and biosphere. These interactions exhibit complex patterns and dynamics, often resulting in uncertainties in climate prediction models.

> *The Earth's climate system cannot be characterized as a linear system with predictable behavior.*

Even minor perturbations in one component can have disproportionately significant effects on other regions, making it challenging to accurately forecast future climate states.

The system is replete with feedback loops, some magnifying changes while others mitigating them. This complexity necessitates intricate modeling efforts to capture and simulate these dependencies.

While observational data is indispensable for model development and evaluation, substantial gaps persist in our comprehension of the Earth system, particularly in data from remote and under-sampled regions. Numerous small-scale processes, such as cloud formation and ocean mixing, are too intricate to be explicitly represented in climate models. Consequently, they are parameterized, employing simplified representations that can introduce uncertainties.

Despite utilizing the most powerful supercomputers, current climate models still encounter limitations in their resolution and the capacity to accurately simulate all pertinent processes. This results in uncertainties in predictions.

The Earth's climate system exhibits natural variability on various time scales. This inherent variability can make it challenging to discern the signal of human-induced climate change from natural fluctuations, particularly over shorter time periods. Extreme weather events, such as heatwaves, droughts, and floods, inherently possess unpredictability to some extent. While climate models can project alterations in the probability and intensity of such events, predicting their precise timing and location remains an intricate task.

The inherent unpredictability of numerous events, including extreme weather phenomena such as prolonged and unexpected heatwaves, droughts, floods, heavy storms, and brushfires, poses a significant challenge to forecasting. While climate models can provide projections of changes in the likelihood and intensity of these events, accurately predicting their precise timing and location remains an ongoing area of research and development.

Human activities, such as greenhouse gas emissions, land use change, and technological advancements, significantly influence the climate. These factors inherently possess inherent uncertainties and pose challenges in accurately predicting their impacts. Furthermore, there are additional influences, including geoengineering techniques. These techniques have the potential to alter the Earth's climate. Consequently, incorporating these uncertainties into current climate models remains an intricate task.

Even with the most powerful supercomputers, current climate models still have limitations in their resolution and the ability to accurately represent all relevant processes. This leads to uncertainties in predictions.

The Role of Machine Learning

At the highest level, machine learning algorithms can be employed to discern patterns in intricate data and enhance the climate model simulations. This enhancement in model accuracy is achieved by identifying patterns in complex data. Furthermore, these techniques enable a more precise quantification of uncertainties associated with the prediction of climate events.

Concurrently, these algorithms may facilitate the development of more efficient climate models, thereby reducing computational costs.

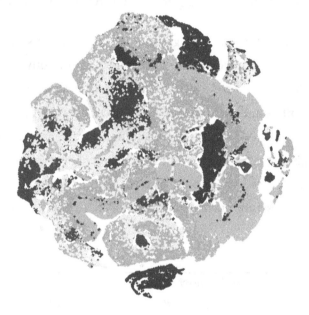

Distributed Stochastic Neighbor Embedding (tSNE) is a technique used to visualize complex, high-dimensional climate datasets by reducing them to a lower dimensional space, typically 2D or 3D, allowing researchers to identify patterns, clusters, and relationships between different climate variables.

Climate Data Sources and Preparation

There are a variety of data sources and specific methods to collect the data properly.

Atmospheric data contains measurements of air temperature, surface temperature, and sea surface temperature labeled as *temperature* type. Rainfall and rain intensity, as well as snowfall and snow depth, are categorized as *precipitation*. *Wind speed and direction* are measured by monitoring the surface wind and upper-level winds.

Air pressure at different altitudes is *atmospheric pressure,* while relative humidity and specific humidity are collected under *humidity.* We also measure dust, smoke, sea salt, and volcanic ash under *aerosols* and carbon dioxide (CO_2), methane (CH_4), water vapor, and nitrous oxide (N_2O) as *greenhouse gas concentrations.*

We collect these data types with surface-based and upper-air stations. Remote sensing data collection happens with satellites like MODIS, Landsat, radar, and LiDAR.

Aircraft can be equipped with various sensors for aerial observations and buoys for oceanic collection with meteorological sensors. In addition, ground-based observation from volunteers completes the picture.

For oceanic data, the types contain changes in coastal elevation by measuring the absolute and relative sea levels. Surface currents and deep-ocean currents create the data type *ocean currents.*

Salinity levels, pH levels, and dissolved inorganic carbon at different depths for *ocean salinity* and *ocean acidification* are added to the data types. Extremely important are changes in the *sea surface temperature, marine heatwaves, and marine life, which measure* the duration and intensity of ocean warming events by abundance or significant distribution changes of marine species.

Oceanic data is collected with floats measuring temperature, salinity, and complemental parameters at various depths (Argo Floats). Moored and drifting oceanographic buoys are equipped with specific sensors to add data as well as research vessels and satellites for altimetry and scatterometry.

Sound waves are used to measure ocean currents.

Cryospheric data comprises Arctic and Antarctic sea ice extent and thickness, changes in the glacier volume, and ice flow. Changes in Greenland and Antarctic ice sheets, the seasonal snow cover on land, and the thawing of permafrost complete this data type. This data type is collected using satellites with radar and optical imagery, GPS, radar altimetry, and ice core drilling. Using aircraft for aerial surveys and photography completes the collection tooling.

For the land surface of the earth, terrestrial data contains air temperature, soil temperature, rainfall, and snowfall. Changes in forest cover, agriculture, and urbanization measurements, as well as the water content of the soil, vegetation greenness, leaf area index, and extent of wildfires, complete the data.

The terrestrial data is collected with ground-based stations equipped with meteorological sensors in collaboration with satellites that provide measurements and aerial photography. Field observations and plant and animal species sampling are essential, including tree ring analysis, to understand and reconstruct past climate conditions.

The next important data type is the biogeochemical, which consists of the CO_2 exchange between the atmosphere and the land/ocean. Methane (CH_4) emissions come from agriculture, wetlands, and various other sources, as well as nitrogen, phosphorus, and other elements—some directly influencing the changes in seawater pH and carbonate chemistry.

Flux towers measure the collection methods for these data types through the exchange of greenhouse gases between the ecosystem and the atmosphere. Stable isotopes are required to trace carbon sources and sinks. The more concise picture comes by developing and running models to simulate biogeochemical processes.

The last data type, at least as important as any other, is the socioeconomic data. The subtypes are population growth and distribution, the type of energy consumption balance between fossil fuels and renewable energy. Industrial output and the gross domestic product, as well as the land usage practices, crop yields and water use, water availability, and water quality, complete this important data type.

Data collection methods for this data type include population surveys and demographic information. Statistical data is also obtained from national and international agencies. Monitoring land use and changes in urbanization is also important, as are surveys and questionnaires that collect human behavior and attitudes. Advanced climate research relies on data sharing and international collaboration. Therefore, data accuracy, consistency, and completeness are crucial before sharing. Combining data from different disciplines and sources requires complex review and preprocessing steps to avoid redundancy and gaps in the collected data.

Machine Learning Concepts in Climate Research

Machine learning has emerged as a potent tool in climate science, providing novel methodologies for analyzing intricate climate data, identifying patterns, and formulating predictions. This comprehensive guide elucidates fundamental ML concepts and their practical applications in climate research, specifically focusing on time series analysis, supervised and unsupervised learning, deep learning, and feature engineering.

Time Series Analysis

Time series analysis is fundamental in climate research, as climate change typically exhibits sequential temporal patterns. By analyzing these patterns, scientists can discern temporal trends and make predictions at various scales. This analysis is particularly valuable for temperature trend analysis spanning different time periods, sea level rise projections, and extreme weather event forecasting. El Niño and La Niña events are closely monitored, and seasonal predictions are made by studying the temporal sequence.

Time series data analysis and decomposition serve as the foundation of climate research. Trend analysis provides insights into potential long-term climate change patterns. Localized anomaly analysis aids in consolidating unexpected variations within the underlying data, enabling the identification of underlying patterns.

ARIMA[2] models predict future values based on past data patterns. When these patterns include seasonal cycles (like temperature variations throughout the year), the SARIMA[3] extension is used – it's essentially ARIMA with added seasonal prediction capabilities.

In simpler terms, ARIMA predicts trends, while SARIMA handles both trends and recurring seasonal patterns.

Complex temporal dependencies show that LSTM[4] is particularly valuable for modeling. There are complex temporal dependencies in the climate system.

---

[2] ARIMA stands for Autoregressive Integrated Moving Average for forecasting of possible future values.

[3] SARIMA Seasonal Autoregressive Integrated Moving Average.

[4] LSTM Long-Short-Temp-Memory network belongs to the recurrent neural network (RNN) algorithms to remember information over time for predictions.

Supervised Learning

Classification and regression are supervised learning algorithms that belong to the classification and regression tasks. In the domain of extreme weather event detection and analysis, cloud type categorization, and the separation of extreme weather patterns, classification methods prove to be highly effective.

These approaches empower meteorologists and climate scientists to gain a deeper comprehension of atmospheric phenomena, thereby augmenting the accuracy of weather forecasting.

Regression algorithms are highly suitable for predicting continuous variables such as rainfall amounts, temperature, and greenhouse gas emissions.

These predictions are fundamental to climate modeling and serve as a basis for policy decisions regarding climate change mitigation strategies. Sea level projection models, in particular, rely heavily on regression techniques to forecast future changes in ocean levels.

Support Vector Machines (SVMs) have found their niche in climate pattern classification tasks. Their effectiveness in handling high-dimensional climate data makes them ideal for analyzing multiple atmospheric variables simultaneously. They are particularly useful for detecting boundaries between different weather systems and climate zones.

Gradient Boosting techniques have demonstrated exceptional accuracy in temperature prediction tasks. These algorithms excel at capturing non-linear relationships between variables, which is crucial in climate science, where interactions between different factors are often complex and interdependent. Their ability to handle missing data scenarios makes them particularly valuable when working with historical climate records that may have gaps or inconsistencies.

Random Forests have become a cornerstone in climate research due to their ability to handle multiple climate variables simultaneously. Their built-in feature importance ranking enables scientists to identify the most significant factors influencing climate patterns. The algorithm's robustness to outliers makes it particularly suitable for climate data, which frequently contains extreme events and anomalous measurements.

Unsupervised Learning Applications

Unsupervised learning techniques have revolutionized our comprehension of climate patterns by enabling the identification of natural groupings and relationships within climate data. Climate zone identification stands as a pivotal application within this domain. Through clustering algorithms, researchers can delineate distinct climate regions and classify ecosystems based on a multitude of environmental variables. This approach has fostered a more nuanced understanding of the evolving climate zones in response to global warming.

Pattern discovery in climate data has been substantially enhanced through unsupervised learning methodologies. EOF/PCA analysis facilitates the identification of dominant modes of climate variability, while self-organizing maps excel at revealing atmospheric circulation patterns. These techniques have proven invaluable for detecting anomalies in climate data and comprehending large-scale climate phenomena.

Dimensionality reduction techniques play a pivotal role in climate science by rendering intricate datasets more manageable and interpretable. Principal Component Analysis assists researchers in developing climate indices that encapsulate fundamental patterns in atmospheric and oceanic variables. T-SNE visualization techniques enable scientists to explore high-dimensional climate data more intuitively, while autoencoders extract pertinent features from complex climate datasets.

Deep Learning in Climate Science

Deep learning has revolutionized climate science by analyzing complex, high-dimensional data. Convolutional Neural Networks (CNNs) revolutionize satellite imagery analysis, identifying cloud patterns, extreme weather, and atmospheric circulation. U-Net and ResNet architectures excel in climate pattern segmentation and classification, respectively. Inception networks analyze features at multiple scales, understanding climate phenomena across dimensions.

Recurrent Neural Networks (RNNs) analyze temporal patterns. Long Short-Term Memory (LSTM) networks excel in long-term temperature forecasting, capturing complex dependencies. Gated Recurrent Units (GRUs) offer computational efficiency while maintaining high accuracy in precipitation and wind pattern prediction. Their application to ocean current modeling provides insights into marine circulation and its impact on climate systems.

Climate Parameters Feature Engineering

Feature engineering in climate science necessitates a profound comprehension of both physical processes and data characteristics. Temperature-related features constitute a fundamental aspect of climate analysis, encompassing not only fundamental measurements, but also derived metrics, such as temperature gradients and heat indices. These parameters must undergo meticulous processing to account for diurnal and seasonal cycles, while Growing Degree Days (GDDs) provide pivotal information for agricultural applications and ecosystem studies.

Precipitation features present unique challenges in feature engineering due to their highly variable nature. Beyond basic rainfall measurements, researchers must consider intensity patterns, drought indices, and accumulated precipitation effects. Snowwater equivalent calculations further complicate the process, necessitating careful consideration of temperature, pressure, and humidity conditions. These features often necessitate specialized processing to account for measurement uncertainties and spatial variability.

Atmospheric features represent some of the most intricate parameters in climate science. Pressure gradients and wind components must be analyzed in three dimensions, while humidity indices require careful consideration of temperature-dependent saturation points. Cloud cover metrics present challenges due to their dynamic nature and difficulty obtaining consistent measurements across diverse observation platforms.

The development of derived features has undergone significant advancement in climate science. Statistical features extend beyond simple averages to encompass rolling statistics that capture temporal variations at multiple scales. These include precisely calculated means and variances, along with indicators designed to capture extreme events and long-term trends. Seasonality indices assist in quantifying changes in annual patterns, while trend indicators track long-term climate evolution.

Spectral features provide pivotal insights into climate periodicities and patterns. Fourier transforms of climate time series unveil dominant frequencies in climate variations, while wavelet coefficients elucidate the temporal evolution of these patterns. Power spectral density analysis facilitates the identification of dominant modes of variability, and cross-spectral analysis elucidates relationships between disparate climate variables.

Spatial features elucidate the geographic aspects of climate patterns. These encompass meticulously constructed gradient maps that reveal spatial variations in climate parameters, complemented by correlation analyses that identify spatial correlations between various regions. Geographic indicators incorporate the influence of latitude, longitude, and elevation on climate patterns, while topographic features account for the effects of the landscape on local climate conditions.

Best Approach and Consideration

Data preprocessing in climate science necessitates rigorous attention to detail. Missing data imputation must account for the spatial and temporal correlations inherent in climate systems, while outlier detection requires careful consideration of extreme events that may represent genuine climate signals rather than measurement errors. Temporal alignment of different data sources presents particular challenges, especially when working with satellite and ground-based measurements. Spatial interpolation must be performed carefully, considering topography and other geographic features that influence climate patterns.

The normalization of climate data necessitates sophisticated approaches that preserve important signals while making the data suitable for machine learning applications. Standard scaling must be applied judiciously, as climate data often exhibits pronounced seasonal patterns and long-term trends that should be preserved.

Robust scaling techniques are particularly pertinent when dealing with climate extremes, while variable transformations must be selected judiciously to maintain physical relationships between different parameters.

Model validation in climate science presents unique challenges due to the spatial and temporal nature of the data. Cross-validation strategies must be adapted to account for temporal dependencies, often employing specialized techniques such as block validation or leave-one-out methods modified for time series. Performance metrics must transcend standard statistical measures to encompass domain-specific considerations, such as capturing extreme events or maintaining physical consistency.

Climate-specific challenges necessitate careful consideration throughout the modeling process. The handling of spatial and temporal dependencies demands sophisticated approaches that maintain physical consistency while capturing intricate relationships. The treatment of climate extremes requires special attention, as these events often carry significant importance despite their relative rarity. Data quality issues must be addressed with meticulous consideration of measurement uncertainties and systematic biases. Incorporating physical constraints ensures that machine learning models produce results consistent with established physical principles, while maintaining scientific interpretability remains paramount for the acceptance and application of model results in climate science.

Applying machine learning in climate research strikes a delicate equilibrium between computational complexity and physical comprehension. Achieving success in this domain necessitates not only technical proficiency in machine learning methodologies, but also a profound understanding of the intricate dynamics of climate systems. By meticulously balancing these considerations, researchers can construct models that advance our comprehension of climate processes while upholding scientific rigor and physical coherence.

The integration of machine learning with climate science remains an evolving process, providing novel insights and capabilities in our pursuit of understanding and forecasting climate change.

Core Concepts and Applications

Global and Regional Temperature Trends

Given the sequential nature and interdependence of regional and global temperature fluctuations, time series analysis techniques are the preferred method for temperature analysis. One commonly employed method is Seasonal Autoregressive Integrated Moving Average (SARIMA), which is particularly effective for temperature analysis due to its ability to accommodate both seasonal patterns and long-term trends. Scientists utilize SARIMA to decompose collected temperature data into seasonal and residual trends.

Accounting for annual temperature cycles is highly advantageous, facilitating the identification of underlying cooling and warming patterns. SARIMA can incorporate external regressors such as CO_2 and solar activity levels, making the data stationary.

The open-source Python library Prophet is specifically designed for time series with seasonal effects and multiple seasonal periods. It effectively handles missing data and outliers in the utilized records. Prophet proves particularly useful for analyzing urban heat islands and identifying temperature trend changepoints. Additionally, uncertainty intervals for prediction can be generated using the library.

The subsequent group of algorithms from the deep learning approach are Long Short-Term Memory (LSTM) Networks. These are designed to capture long-term dependencies by simultaneously processing multiple input variables, such as temperature, humidity, and pressure.

Climate data exhibits non-linear relationships with all natural events, making LSTM networks highly suitable for handling such data. Furthermore, LSTM networks can be configured as an encoder-decoder approach for sequential predictions. Climate research often involves dealing with intricate patterns, which LSTM networks can effectively address.

Gaussian Process Regression provides uncertainty estimates and is an effective approach for analyzing sparse and unevenly sampled temperature data. If prior knowledge about climate physics is available, this method can incorporate it. One of its primary contributions is the interpolation of missing temperature data and the capability to provide confidence intervals for predictions.

For ensemble methods that combine multiple decision trees, handling non-linear relationships and interactions between variables, while simultaneously incorporating multiple predictors such as aerosols, solar activities, and greenhouse gases, and providing feature importance ranking to identify key components of temperature changes, Random Forest algorithms are employed. These algorithms simultaneously exhibit resistance to overfitting and are crucial for handling noisy data.

For modeling physical climate systems, neural ordinary differential equations (neural ODEs) constitute a sophisticated amalgamation of differential equations and neural networks. This approach is particularly well-suited due to its capacity to accommodate irregular sampling intervals. By providing continuous-time predictions and incorporating physical constraints from climate science, neural ODEs offer a robust framework for comprehending intricate climate dynamics.

Bayesian Neural Networks (BNNs) introduce probabilistic rigor to temperature analysis by providing uncertainty estimates for predictions. They excel at incorporating prior knowledge about climate physics while handling noisy and incomplete temperature records. This makes them particularly valuable for quantifying prediction confidence and detecting anomalous temperature patterns that may indicate significant climate events.

Transformer Networks with Attention mechanisms excel at capturing long-range dependencies in temperature data. These networks process global temperature patterns across multiple locations simultaneously, making them invaluable for analyzing spatial-temporal relationships. Their ability to handle multiple input variables effectively makes them particularly suitable for integrating diverse data sources, including both satellite data and ground measurements.

For feature engineering, Empirical Mode Decomposition (EMD) emerges as a potent tool for decomposing temperature signals into intrinsic mode functions. This technique effectively separates natural variability from anthropogenic trends and proves particularly effective for analyzing multi-decadal oscillations. EMD's capability to handle non-linear and non-stationary data renders it exceptionally suitable for analyzing intricate climate cycles.

Wavelet Analysis provides a sophisticated approach to examining temperature patterns at various time scales. This method identifies periodic components in temperature data and detects regime shifts. Its distinguishing feature lies in separating local and global temperature patterns, making it particularly effective for analyzing extreme weather events within their broader climate context.

Transfer Learning Approaches leverage pre-trained models from related climate tasks, which is particularly valuable for regions with limited data. By incorporating knowledge from existing climate models, these approaches reduce computational requirements while enhancing prediction accuracy for specific regions. This makes them particularly useful for developing targeted regional climate assessments.

The validation process for these models necessitates specialized Cross-Validation techniques for time series data. This involves meticulous time-based splitting of training and validation data to account for temporal dependencies and prevent future information leakage. Such validation is crucial for assessing model performance across diverse time periods and ensuring reliable temperature predictions.

Uncertainty Quantification plays a pivotal role in climate analysis, employing bootstrap methods for confidence intervals and ensemble predictions for uncertainty estimates. Monte Carlo methods facilitate error propagation, making these techniques indispensable for policy-relevant predictions. This comprehensive approach to uncertainty identification highlights areas where additional data collection may be necessary.

Multi-scale Analysis combines local and global temperature patterns while accommodating different temporal resolutions.

This approach accounts for spatial correlations and proves particularly useful for regional climate assessments. The capability to analyze climate patterns at multiple scales renders it an invaluable tool for adaptation planning and policy development.

Ensemble methods and hybrid approaches frequently employ stacked models that combine predictions from multiple algorithms, including both statistical and machine learning methods. This approach reduces prediction variance and effectively addresses various aspects of temperature variation compared to single-model approaches. The resulting predictions tend to be more robust and reliable.

Implementing these methods requires careful consideration of several factors, including data quality and availability, temporal and spatial resolution requirements, computational resources, specific analysis objectives, and required prediction horizons. For operational use, organizations must consider regular model retraining with new data, efficient handling of missing or erroneous measurements, computational efficiency requirements, interpretability needs, and integration capabilities with existing climate models.

The field of machine learning for climate analysis continues to evolve rapidly, with new methods and approaches being developed regularly. Success in this domain often comes from thoughtfully combining multiple approaches while incorporating domain knowledge from climate science. This integrated approach ensures that analyses remain both scientifically rigorous and practically useful for understanding and predicting climate trends.

Heat Waves: Temperature Anomalies and Urban Heat Islands

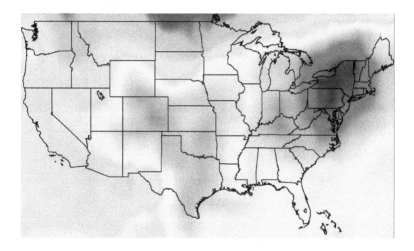

Individuals exposed to heat waves and urban heat islands experience substantial physiological stress. In urban environments, where temperatures can be 4-7°C higher than surrounding rural areas, people frequently encounter heightened discomfort due to the synergistic effects of elevated temperatures, humidity, and restricted airflow between buildings. The human body's capacity to regulate core temperature through sweating is compromised when humidity is elevated and air circulation is limited. This can result in symptoms such as fatigue, dizziness, and, in severe cases, heat exhaustion or heat stroke. Vulnerable populations, including elderly individuals, young children, and those with pre-existing health conditions, are particularly susceptible to these adverse effects.

Machine learning algorithms have significantly advanced our comprehension and forecasting of heat waves and urban heat islands by processing vast amounts of environmental data and identifying intricate patterns. Deep Learning Neural Networks (DLNNs) have emerged as potent tools for temperature prediction and heat wave forecasting. These networks excel at capturing non-linear relationships between diverse environmental parameters, such as solar radiation, wind patterns, humidity, and urban infrastructure characteristics. Convolutional Neural Networks (CNNs) have demonstrated exceptional efficacy in processing satellite imagery and thermal maps to discern urban heat island patterns and their temporal evolution.

Support Vector Machines (SVMs) have demonstrated remarkable success in classifying diverse urban microclimates and predicting extreme temperature events. Their efficacy stems from their ability to handle high-dimensional data and discern intricate decision boundaries between various temperature regimes. Researchers have employed SVMs to analyze the correlation between urban morphology and temperature distribution, thereby identifying areas most susceptible to extreme heat events.

Random Forest algorithms have exhibited exceptional success in elucidating the contributing factors to urban heat island formation. These ensemble learning methodologies effectively manage multiple variables simultaneously, including building density, vegetation cover, surface materials, and anthropogenic heat sources. The distinguishing advantage of Random Forests lies in their capacity to rank feature importance, enabling urban planners to identify the most influential factors influencing local temperature variations.

At its fundamental basis, a random forest operates by constructing numerous decision trees, employing an ingenious strategy. Unlike conventional approaches that permit each tree to access all available information, the algorithm ensures that each tree develops its expertise based on a random selection of data and features. This methodology aligns with the concept of diverse experts specializing in various aspects of a problem, resulting in more comprehensive decisions when their insights are combined.

To illustrate this concept, consider the process of predicting whether an individual will enjoy a movie. In this scenario, each decision tree within the forest develops its prediction based on various aspects of the movie-watching experience. One tree may prioritize the genre and the lead actor's previous performances, while another examines the director's track record and the movie's budget. A third tree could concentrate on factors such as the movie's length and release timing. Each tree forms its own opinion based on these partial views of the data.

The true power of random forests becomes evident when these individual predictions are combined. By adopting a democratic approach, essentially a collective decision-making process involving all the trees, the algorithm produces a final prediction that benefits from multiple perspectives. This collective decision-making process tends to be more reliable than relying on any single decision tree, as it mitigates individual biases and captures intricate patterns in the data.

The combination of randomness in feature selection and data sampling, coupled with the aggregation of multiple predictions, renders random forests particularly robust and accurate. They excel at handling intricate relationships in data while mitigating the pitfalls of excessive reliance on any single aspect of the available information. Consequently, the resulting algorithm is a potent and versatile tool that has demonstrated its efficacy in numerous practical applications.

Recurrent Neural Networks (RNNs), particularly Long Short-Term Memory (LSTM) networks, have proven invaluable in capturing the temporal dynamics of heat wave development. These networks can process sequential temperature data and account for seasonal patterns, rendering them particularly suited for long-term temperature trend analysis and heat wave prediction.

Recurrent Neural Networks (RNNs) are neural networks that process information sequentially, akin to reading a story. Each word's significance extends beyond its immediate context, as it is influenced by the preceding words. This "memory" mechanism is achieved through internal connections that propagate information backward within the network, enabling continuous processing of context-aware data.

To illustrate this concept with a climate-related example, we consider the task of predicting temperature patterns. Meteorologists recognize that today's temperature is not merely a standalone value but is intricately linked to yesterday's conditions, seasonal variations, and longer-term climate trends. An RNN adopts a similar approach to address this problem.

For instance, when tracking daily temperature variations in a coastal city, the RNN processes this data by simultaneously maintaining awareness of multiple timescales. At each step, it considers the immediate temperature reading, temperature patterns from the past few days, longer-term seasonal variations, and the interplay between ocean temperatures and land temperatures.

The network's memory cells retain essential information regarding these patterns. For instance, when analyzing summer temperature spikes, the RNN does not merely perceive a high temperature in isolation; it comprehends this in the context of seasonal progression, recent weather patterns, and historical trends. This contextual understanding enables it to discern subtle patterns, such as the potential lag between ocean and land temperature changes or the influence of morning temperatures on afternoon peaks.

What distinguishes RNNs, particularly in climate analysis, is their capacity to simultaneously learn and recognize patterns across diverse time scales. They can detect short-term weather patterns while simultaneously maintaining awareness of longer-term climate trends. This multifaceted capability renders them highly valuable for comprehending intricate climate phenomena where alterations in one time period can have cascading effects much later.

For instance, when studying the effects of El Niño events, recurrent neural networks (RNNs) can track how initial changes in ocean temperatures lead to cascading effects in atmospheric conditions over weeks and months. The network's memory mechanism enables it to retain relevant information about these evolving patterns, facilitating the identification of connections between early indicators and subsequent climate impacts that may not be apparent in the raw data.

This combination of sequential processing and memory retention renders RNNs particularly well-suited for climate science applications, where comprehending the interconnectedness of environmental systems is paramount for accurate analysis and prediction.

Long Short-Term Memory (LSTM) networks represent a significant advancement in sequence modeling, addressing fundamental limitations of traditional Recurrent Neural Networks (RNNs) through an ingeniously designed memory management system. Unlike their simpler RNN counterparts, LSTMs possess sophisticated mechanisms to control information flow, enabling them to effectively capture both long-term patterns and short-term fluctuations in complex sequential data.

LSTMs have specifically demonstrated efficacy in forecasting the duration and intensity of heat waves, providing invaluable information for public health planning.

In the realm of climate analysis, Long Short-Term Memory (LSTM) networks exhibit their true potential by possessing the capability to maintain and selectively update multiple layers of temporal information. Consider the challenge of comprehending global temperature patterns. An LSTM can simultaneously monitor daily variations, seasonal cycles, and long-term climate trends through its distinctive cell state and gating mechanisms.

The cell state within an LSTM functions akin to a meticulously organized climate record, wherein information traverses a series of gates that dictate retaining, modifying, or discarding data. Envision tracking Arctic sea ice extent: the forget gate may acquire the ability to discard daily fluctuations while preserving pivotal seasonal transition points. The input gate judiciously integrates novel measurements with existing patterns, while the output gate determines which aspects of accumulated knowledge are most pertinent for current predictions.

This sophisticated memory management becomes particularly valuable when analyzing climate phenomena that operate on multiple time scales. For instance, when studying the relationship between ocean temperatures and atmospheric patterns, an LSTM can maintain information about slow-changing ocean heat content while simultaneously processing rapid atmospheric fluctuations. The network's ability to selectively preserve information enables it to recognize how subtle changes in ocean temperatures may influence weather patterns weeks or months later.

What distinguishes LSTMs in climate applications is their ability to address the "vanishing gradient problem" that affects simpler recurrent neural networks (RNNs). This enables them to effectively learn relationships between events separated by extended time intervals, which is crucial for comprehending climate patterns that may manifest over months or years. For example, when analyzing the development of drought conditions, an LSTM can retain relevant information about precipitation patterns, soil moisture levels, and temperature trends over extended periods, allowing it to identify subtle precursors to drought conditions that simpler models may overlook.

The gates within an LSTM also serve as a form of attention mechanism, automatically identifying which aspects of historical data are most pertinent for current predictions. In climate modeling, this may entail learning to prioritize specific seasonal patterns that historically precede extreme weather events, while simultaneously maintaining awareness of longer-term climate trends that may influence these relationships.

The intricate interplay between short-term and long-term memory renders LSTMs particularly valuable for climate science applications, where comprehending intricate, multi-scale temporal connections is paramount for accurate prediction and analysis of environmental phenomena.

The last in this group, Gradient Boosting Machines (GBMs), represents an elegant approach to machine learning that builds powerful predictive models through an iterative process of learning from mistakes. Unlike random forests, which build many trees in parallel and then combine their predictions, gradient boosting constructs its model sequentially, with each new addition specifically designed to correct the errors of its predecessors.

GBMs, such as LightGBM and XGBoost, have demonstrated remarkable performance in integrating multiple environmental parameters to forecast local temperature variations. These algorithms manage diverse data types and can effectively process continuous variables (e.g., temperature and humidity) and categorical variables (e.g., land use types and building materials). GBMs have proven particularly valuable in developing early warning systems for heat waves by integrating historical temperature data with real-time measurements.

The Gradient Boosting Decision Tree (GBDT) can be conceptualized as an expert team that evolves through meticulous recruitment. Each new member is deliberately selected to address the existing team's deficiencies. In climate analysis, this process might commence with a rudimentary model that encapsulates fundamental temperature patterns. If this initial model consistently underpredicts temperature surges during heat waves, the subsequent addition to the ensemble would specifically concentrate on enhancing the model's ability to predict these extreme events.

The term "gradient" in gradient boosting refers to the mathematical procedure of quantifying and systematically reducing prediction errors. This process is akin to how a climate scientist iteratively refines their comprehension of a complex weather system, directing their attention to the aspects of the phenomenon that are least well-understood. Each new component of the model is trained to predict the residual errors, the disparity between actual and predicted values, from the preceding iterations.

A practical example in climate modeling is predicting regional rainfall patterns. The initial model may capture fundamental seasonal trends but struggle with predicting sudden downpours. The subsequent model would then concentrate specifically on these precipitation extremes. A third model might address the intricate interactions between temperature and humidity that the preceding models overlooked. Each iteration adds another layer of sophistication, with the final prediction integrating all these specialized components.

What distinguishes gradient boosting, particularly from other methods, is its ability to automatically identify and focus on the most challenging aspects of the prediction task. In climate applications, this may entail automatically discovering complex relationships between multiple variables, such as how wind patterns, humidity, and temperature interact to produce specific weather conditions. The algorithm naturally allocates more computational resources to comprehending these intricate patterns while dedicating less effort to relationships that are already adequately captured.

The sequential nature of gradient boosting inherently provides a form of regularization. Each new addition to the model is typically constrained to make relatively simple predictions, preventing any single component from becoming overly complex or overfitted to the training data. This results in powerful and robust models capable of capturing intricate patterns while maintaining a good generalization to new situations.

This combination of iterative refinement, focused error correction, and controlled complexity makes gradient boosting machines particularly effective for climate-related predictions where the interaction between different variables can be subtle and complex, yet critically important for accurate forecasting.

Machine learning has also been instrumental in developing hybrid models that combine physical understanding with data-driven approaches. For instance, physics-informed neural networks (PINNs) incorporate known physical laws and constraints into their learning process, improving the accuracy of temperature predictions while ensuring physically consistent results. These hybrid approaches have been particularly valuable in understanding the complex interactions between urban infrastructure and local climate.

Applying models trained on well-studied urban areas to cities with limited historical data transfer learning techniques is very helpful.

This approach has been particularly valuable for developing regions where extensive temperature monitoring networks may not be available. By adapting pre-trained models to local conditions, researchers can provide reasonable predictions of urban heat island effects and heat wave patterns.

Subsequently, unsupervised learning methods, particularly clustering algorithms such as DBSCAN and K-means, have facilitated the identification of distinct thermal patterns within urban environments. These algorithms can automatically discern areas exhibiting similar temperature profiles, enabling urban planners to pinpoint urban heat island hotspots and prioritize mitigation strategies. Dimensionality reduction techniques, such as Principal Component Analysis (PCA), play a crucial role in comprehending the intricate relationships between various factors contributing to urban heat islands.

The seamless integration of machine learning with Internet of Things (IoT) sensor networks has revolutionized urban temperature monitoring and prediction. Edge computing implementations of lightweight machine learning models facilitate the rapid processing of sensor data, promptly triggering alerts in the event of hazardous heat conditions. This integration has proven particularly valuable in developing smart city applications capable of dynamically responding to evolving temperature scenarios.

Currently, reinforcement learning approaches have emerged as tools for optimizing urban design to mitigate heat island effects.

These methods can simulate various urban development scenarios and their impact on local temperature patterns, enabling planners to make informed decisions regarding building placement, green space allocation, and material selection. Although still in its nascent stages, reinforcement learning demonstrates the potential for developing adaptive urban planning strategies that adapt to changing climate conditions. Enhancing the interpretability of models remains a key challenge for the future of machine learning algorithms.

Extreme Weather and Precipitation

Rainfall Patterns: Drought Prediction and Flood Risk Assessment

These points encompass crucial aspects significantly impacting humans globally.

ML and AI techniques are employed in analyzing and predicting precipitation patterns, drought conditions, and flood risks. This domain presents a critical area where AI has made substantial contributions to environmental and disaster management.

In the context of rainfall pattern analysis, deep learning architectures have emerged as potent tools for processing the intricate, multidimensional data involved in precipitation forecasting. Convolutional Neural Networks (CNNs) have demonstrated exceptional efficacy in analyzing spatial rainfall patterns derived from satellite imagery and weather radar data. These networks excel at identifying features within gridded precipitation data, simultaneously processing multiple layers of meteorological information to discern patterns that may elude conventional statistical methods.

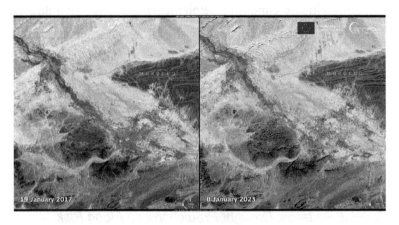

Centuries-old oases in Morocco under drought.

Recurrent Neural Networks (RNNs), particularly Long Short-Term Memory (LSTM) networks, as outlined in the previous discussion, have become instrumental in capturing temporal dependencies in rainfall patterns. These architectures are particularly well-suited for analyzing time series data, enabling them to identify seasonal patterns, trends, and cyclical behaviors in precipitation records. The ability of LSTMs to maintain information about long-term dependencies while simultaneously being sensitive to recent changes makes them particularly valuable for predicting rainfall patterns influenced by both immediate weather conditions and longer-term climate cycles.

Ensemble learning methods have demonstrated remarkable success in predicting droughts. Random Forests and Gradient Boosting Machines (GBM) are commonly employed to process multiple drought indicators simultaneously. These algorithms can effectively combine various input parameters, such as precipitation indices, soil moisture levels, vegetation health indicators, and temperature anomalies, to produce robust drought forecasts. The advantage of ensemble methods lies in their ability to address the inherent uncertainty in drought prediction by aggregating multiple predictive models.

Traditional machine learning algorithms, such as Support Vector Machines (SVMs), have also proven valuable in drought forecasting, particularly when dealing with the non-linear relationships between different environmental variables. SVMs excel at identifying the optimal boundaries between different drought severity classes, making them useful for categorical drought prediction. When combined with kernel methods, SVMs can effectively capture complex patterns in high-dimensional environmental data.

In the context of drought prediction, hybrid models are employed that combine various machine learning approaches. For instance, combining convolutional neural networks (CNNs) to process satellite imagery for vegetation health assessment with long short-term memory (LSTM) networks to analyze temporal patterns in precipitation and temperature data outperforms single-algorithm solutions. This is because hybrid models can capture both spatial and temporal aspects of drought development.

In flood risk assessment, AI techniques have advanced to address the unique challenges of flood prediction, including the need to process both static and dynamic risk factors. Deep learning models have been developed to analyze digital elevation models (DEMs), land use patterns, and soil characteristics in conjunction with real-time precipitation data and river gauge measurements. These models often incorporate attention mechanisms to prioritize the most relevant features for different types of flooding events, such as river overflow, coastal surge, or intense rainfall.

Bayesian networks have proven particularly valuable in flood risk assessment due to their ability to handle uncertainty and incorporate expert knowledge. These probabilistic models can combine historical flood data with physical parameters and human expertise to produce probability distributions of flood risk under various scenarios. The Bayesian approach is especially useful in areas where historical data might be limited, as it can effectively combine sparse data with expert knowledge to produce meaningful risk assessments.

Geographic Information System (GIS) integration with machine learning has revolutionized spatial aspects of flood risk assessment. Neural networks designed to process geospatial data can analyze multiple layers of geographic information, including topography, drainage patterns, and infrastructure locations, to identify areas at the highest risk of flooding. These models often incorporate Graph Neural Networks (GNNs) to process the interconnected nature of river networks and drainage systems.

Modern flood risk assessment systems often employ reinforcement learning techniques to optimize emergency response strategies. These algorithms can learn from historical flood events to recommend optimal flood control measures, such as the timing of reservoir releases or the deployment of flood barriers. The reinforcement learning approach is particularly valuable because it can adapt to changing conditions and learn from both successes and failures in flood management.

Transfer learning has become increasingly important in all these applications, particularly for regions with limited historical data. Models trained on data-rich regions can be adapted to predict patterns in areas with sparse data, making AI-based prediction more globally accessible. This approach has been particularly successful in developing countries where historical weather records might be limited.

The integration of physical models with machine learning techniques has led to the development of hybrid systems that combine the best aspects of both approaches. These Physics-Informed Neural Networks (PINNs) incorporate known physical laws and constraints into the learning process, ensuring that predictions remain physically plausible while benefiting from the pattern-recognition capabilities of machine learning.

As computational power and data availability continue to increase, these AI techniques become more sophisticated and accurate. The development of explainable AI (XAI) methods is particularly important in this field, as stakeholders need to understand and trust the predictions made by these systems. Techniques such as LIME (Local Interpretable Model-agnostic Explanations) and SHAP (SHapley Additive exPlanations) are being increasingly incorporated into environmental prediction systems to provide interpretable results.

Forecasting Extreme Weather: Uncertainty Quantification

Detailed explanations of extreme weather forecasting techniques focusing on uncertainty quantification methods in climate measurements and predictions are discussed below.

Extreme weather forecasting represents one of the most challenging aspects of meteorological prediction, requiring sophisticated approaches to handle both the complexity of the underlying phenomena and the inherent uncertainties in measurement and prediction. The foundation of modern extreme weather forecasting lies in Ensemble Prediction Systems (EPS) that generate multiple forecasts by varying initial conditions and model parameters to capture the range of possible outcomes.

The quantification of uncertainty begins at the measurement level, where multiple sources of error must be considered. Instrumental uncertainty arises from the limitations and calibration of weather sensors, ranging from ground-based thermometers and rain gauges to satellite-based remote sensing instruments. Each measurement comes with its own error distribution, typically characterized through rigorous calibration processes and expressed as confidence intervals. For satellite measurements, uncertainty quantification must account for atmospheric interference, sensor degradation over time, and variations in viewing angles.

In the context of extreme weather prediction, Bayesian hierarchical models have emerged as a powerful framework for incorporating these measurement uncertainties into the forecasting process. These models explicitly represent different levels of uncertainty, from measurement error to model parameter uncertainty to natural variability. The Bayesian approach allows for integrating prior knowledge about weather patterns with new observations, producing posterior probability distributions that capture the full range of uncertainty in predictions.

Monte Carlo methods play a crucial role in quantifying uncertainties in extreme weather forecasting. Particularly, Markov Chain Monte Carlo (MCMC) simulations are used to sample from the complex probability distributions that characterize extreme weather events. These methods generate thousands of possible scenarios, each representing a plausible evolution of weather conditions given the uncertainties in initial conditions and model parameters. The resulting ensemble of predictions provides a basis for probabilistic forecasts and risk assessments.

The challenge of uncertainty quantification becomes particularly acute when dealing with rare events where historical data may be limited. Extreme Value Theory (EVT) provides a mathematical framework for extrapolating beyond observed data to estimate the probability of unprecedented extreme events. However, the application of EVT must carefully account for uncertainties in the tail behavior of probability distributions. Bootstrap methods are often employed to estimate confidence intervals for extreme value parameters, providing a measure of the uncertainty in predictions of rare events.

Multi-model ensembles represent another crucial approach to uncertainty quantification in extreme weather forecasting. By combining predictions from different models, each with its own strengths and weaknesses, forecasters can better understand the structural uncertainty inherent in weather prediction. Bayesian Model Averaging (BMA) provides a formal framework for combining these predictions while accounting for the relative performance and reliability of different models. The weights assigned to different models in the ensemble can themselves be treated as uncertain parameters, leading to more robust predictions.

The spatial aspect of uncertainty quantification presents unique challenges in extreme weather forecasting. Geostatistical methods, particularly kriging and its variants, are used to quantify spatial uncertainty in weather patterns. These methods account for spatial correlation structures while providing prediction intervals that vary across geographic regions. Copula-based approaches have proven valuable for modeling spatial dependencies in extreme events, allowing for a more accurate representation of how uncertainties propagate across regions.

When considering longer time scales, the quantification of uncertainty must account for climate change impacts on extreme weather patterns. This introduces additional layers of uncertainty related to future greenhouse gas emissions, climate sensitivity, and regional climate responses. Hierarchical uncertainty quantification frameworks have been developed to explicitly represent these different sources of uncertainty, from scenario uncertainty to model response uncertainty to internal variability.

The verification of uncertainty estimates presents its own challenges. Proper scoring rules, such as the Continuous Ranked Probability Score (CRPS), are used to assess the quality of probabilistic forecasts. These metrics evaluate both the accuracy of the central prediction and the calibration of the uncertainty estimates. Reliability diagrams and spread-skill relationships provide additional tools for assessing whether predicted uncertainties accurately reflect observed forecast errors.

Data assimilation methods, particularly ensemble Kalman filters and particle filters, play a crucial role in updating uncertainty estimates as new observations become available. These methods provide a formal framework for combining prior forecasts with new measurements, accounting for uncertainties in both the model predictions and the observations. The resulting posterior distributions provide updated estimates of both the current weather state and the associated uncertainties.

Machine learning approaches to uncertainty quantification have gained prominence in recent years. Deep probabilistic models, such as Bayesian neural networks and Gaussian process models, can learn complex uncertainty patterns from historical data. These models can capture both aleatoric uncertainty (inherent variability in the weather system) and epistemic uncertainty (uncertainty due to limited knowledge or data). Techniques like dropout sampling and ensemble deep learning provide practical approaches to estimating prediction uncertainties in complex neural network models.

The communication of uncertainty in extreme weather forecasts remains a critical challenge. Visualization techniques have been developed to effectively convey probabilistic forecasts to both technical and non-technical audiences. These include probability of exceedance maps, ensemble spaghetti plots, and fan charts that show the evolution of uncertainty over time. The development of these communication tools must balance the need for scientific accuracy with the requirement for clear, actionable information for decision-makers.

The integration of uncertainty quantification into operational forecasting systems requires careful attention to computational efficiency. Approximate Bayesian Computation (ABC) methods and variational inference techniques provide computationally tractable approaches to uncertainty estimation in large-scale forecasting systems. These methods allow for real-time updating of uncertainty estimates as new observations become available, making them practical for operational use in extreme weather prediction.

Sea Levels: Ocean Dynamics

Climate change poses significant challenges to our oceans and coastlines, requiring sophisticated modeling and prediction approaches. ML and AI methods have emerged as powerful tools to enhance our understanding and prediction capabilities in oceanography and coastal science. This document explores the key ML/AI approaches that can be applied to three interconnected areas: ocean current modeling, sea level changes, and coastal erosion.

Ocean Current Modeling, Sea Level Change, and Coastal Erosion

Deep learning architectures, such as convolutional neural networks (CNNs), long short-term memory (LSTM) networks, and ensemble learning Random Forests, have been extensively explained in detail and have demonstrated substantial potential in modeling complex ocean currents and their temporal evolution. Their ability to identify mesoscale eddies and current boundaries has proven invaluable in oceanographic research.

Furthermore, deep learning architectures can effectively predict temporal dependencies in ocean current patterns and capture current variations across diverse time scales. Their architectural design facilitates the acquisition and retention of long-term patterns in ocean circulation, making them particularly useful for comprehending seasonal and annual variations in ocean currents. The interaction between various oceanic processes provides an extended perspective and a deeper comprehension of oceanic dynamics.

Physics-Informed Neural Networks (PINNs) represent a revolutionary fusion between traditional neural networks and fundamental physical principles. This integration holds particular significance in climate science, where physical principles govern intricate environmental systems. Unlike conventional neural networks that solely rely on data, PINNs incorporate our knowledge of physical laws directly into their learning process. This hybrid approach strikes a balance between empirical observations and theoretical constraints.

In climate analysis, PINNs excel by integrating fundamental physical principles such as conservation of energy, mass, and momentum into their architectural design. Consider studying ocean currents: a traditional neural network might merely identify patterns from observed data. However, a PINN ensures that its predictions adhere to the Navier-Stokes equations governing fluid dynamics. This integration enables the network to not only discern patterns but also discover physically meaningful patterns that align with our understanding of how ocean systems actually operate.

The efficacy of PINNs becomes particularly evident when confronted with sparse or noisy climate data. Consider the challenge of modeling atmospheric convection patterns: while traditional models may struggle with limited observation points, a PINN can fill in the gaps by leveraging physical laws. The network learns to generate predictions that not only align with available observations, but also adhere to physical constraints such as thermodynamic principles and atmospheric fluid dynamics.

PINNs are particularly valuable in climate science because they can simultaneously handle multiple physical constraints. When modeling climate systems, they can incorporate radiative transfer equations, conservation laws, and chemical reaction dynamics all at once. This results in a sophisticated model that encapsulates the intricate interplay between various aspects of the climate system while ensuring physical consistency across all predictions.

The learning process in PINNs is particularly elegant: the network's loss function integrates both traditional and physics-based data-fitting terms. For instance, when studying sea ice dynamics, the network simultaneously attempts to match observed ice coverage patterns while ensuring its predictions adhere to heat transfer equations and ice formation physics. This dual optimization leads to solutions that are both data-driven and physically sound.

One of the most powerful aspects of PINNs in climate applications is their ability to discover hidden physical parameters. When studying climate feedback loops, a PINN might not only predict system behavior, but also help identify underlying physical constants or relationships that weren't explicitly known. This makes them valuable tools for predicting and advancing our theoretical understanding of climate systems.

Furthermore, PINNs excel at handling multiscale phenomena common in climate science. They can simultaneously capture large-scale atmospheric circulation patterns while respecting small-scale physical processes like cloud formation or aerosol interactions. This multiscale capability, combined with physical consistency, makes them particularly suited for studying complex climate phenomena where processes at different scales interact in physically constrained ways.

The integration of deep learning with physical principles represents a substantial advancement in climate modeling, providing a pathway to fabricate more reliable and physically consistent predictions while preserving the adaptability and pattern-recognition capabilities that render neural networks so formidable.

The fusion of Physics-Informed Neural Networks (PINNs) with Transformer architectures represents a novel convergence in contemporary climate science modeling. This hybrid methodology merges the physical rigor of PINNs with the potent sequential learning capabilities of Transformers, thereby constructing a sophisticated framework capable of simultaneously adhering to fundamental physical principles while capturing intricate temporal dependencies within climate systems.

In the context of global climate patterns, the Transformer's self-attention mechanism enables it to discern relationships between events separated by varying time scales, ranging from daily weather fluctuations to seasonal patterns and long-term climate trends. Concurrently, the PINN component ensures that these discovered relationships are consistent with fundamental physical principles, such as conservation laws and thermodynamic constraints. This fusion of approaches results in a modeling framework that is both data-aware and physically consistent.

The attention mechanism of Transformers demonstrates remarkable efficacy when integrated with physical constraints. In analyzing atmospheric patterns, the model can acquire the ability to focus on physically pertinent relationships that span both space and time. For instance, when investigating the impact of Arctic sea ice on global weather patterns, the Transformer component may discern crucial long-range dependencies, while the PINN framework guarantees that these relationships adhere to the physical laws governing heat transfer and fluid dynamics.

This combined architecture excels in managing the multiscale nature of climate systems. The Transformer's capacity to process sequences of any length harmonizes with the PINN's physical constraints, enabling the model to capture both immediate cause-effect relationships and long-term climate evolution patterns while maintaining physical consistency. For instance, when modeling ocean-atmosphere interactions, the system can simultaneously monitor rapid atmospheric changes and slower oceanic processes, ensuring that their interactions comply with both physical laws and observed temporal patterns.

The learning process within this hybrid system becomes particularly sophisticated. The Transformer's attention weights facilitate the identification of the most relevant historical data elements for current predictions, while the PINN's physics-based loss terms ensure that these predictions remain physically plausible. This fusion results in a powerful synthesis where the model not only learns from data patterns, but also from our fundamental understanding of climate physics.

What makes this combination particularly valuable is its capacity to manage uncertainty and missing data. In regions or time periods with sparse observations, the physical constraints imposed by the PINN component serve as a guiding mechanism for predictions, while the Transformer's attention mechanism facilitates the identification of pertinent patterns from better-observed regions or time periods. This amalgamation results in a resilient system capable of generating physically relevant predictions even in data-deficient scenarios.

Furthermore, this hybrid approach excels in uncovering intricate physical relationships that may be overlooked by either approach independently. The Transformer's attention patterns can illuminate unexpected connections within the data, while the PINN framework facilitates interpreting these connections in the context of physical principles. This convergence can lead to novel insights into climate mechanisms, potentially identifying previously unknown feedback loops or relationships within the climate system.

The culmination of this endeavor is a potent modeling framework that harmoniously integrates the strengths of both approaches: the physical rigor and theoretical consistency of PINNs with the sophisticated pattern recognition and temporal learning capabilities of Transformers. This synthesis heralds a future where climate models can simultaneously leverage our profound understanding of climate systems while capturing the intricate, long-range dependencies that characterize global climate patterns.

Bayesian Neural Networks introduce sophisticated uncertainty quantification to sea level predictions. Their probabilistic approach effectively handles noisy measurements and provides valuable input for risk assessment in coastal planning. These networks enable probabilistic forecasting of sea level changes, which is crucial for adaptive management strategies.

Specifically for coastal erosion analysis, computer vision solutions are powerful tools. Semantic segmentation techniques have revolutionized the analysis of satellite and aerial imagery, enabling precise identification of coastline changes over time. These methods excel at detecting vulnerable coastal areas and monitoring beach morphology changes with unprecedented accuracy.

Object detection networks have advanced our ability to track coastal features and infrastructure. These systems enable continuous monitoring of protective structures and can detect early signs of erosion, allowing for timely intervention. Their capability to identify specific coastal features has enhanced our understanding of erosion processes and improved early warning systems.

RNNs and advanced multi-modal solutions have revolutionized the methodology employed in coastal erosion investigation. These systems seamlessly integrate satellite imagery, weather data, and ocean measurements to provide a holistic environmental monitoring framework. Their capacity to process both visual and numerical data across diverse temporal and spatial scales has substantially enhanced prediction accuracy through sophisticated data fusion techniques.

Modern ML systems have experienced significant advancements in their ability to integrate diverse data sources. Contemporary approaches integrate satellite data with in-situ measurements, while simultaneously incorporating outputs from weather and climate models. These systems also incorporate historical records and expert knowledge, thereby enabling truly comprehensive environmental monitoring that considers all available information sources.

Developing interpretable models has become paramount for practical applications in climate science. Layer-wise Relevance Propagation techniques have enhanced our ability to comprehend neural network decisions. SHAP values provide detailed insights into feature importance, while attention mechanisms assist in identifying key patterns within the data. Advanced visualization techniques have rendered model interpretation more accessible to non-technical stakeholders.

Ocean Acidification: Marine Ecosystem Impact

The ocean chemistry and pH levels are undergoing significant variations across various regions and depths. These changes have cascading effects throughout the marine ecosystem. ML and AI methods assist in analyzing vast amounts of satellite imagery and oceanic sensor data, enabling the identification of patterns and correlations that may be overlooked by conventional statistical methods.

As previously mentioned in other areas, recurrent neural networks (RNNs) and long short-term memory (LSTM) models prove particularly valuable for time series forecasting of acidification trends. These models can capture long-term dependencies and seasonal variations in ocean chemistry. Notably, these models integrate multiple data streams, such as temperature, salinity, dissolved carbon dioxide, and atmospheric carbon dioxide levels, to predict future acidification scenarios with enhanced accuracy.

AI-driven ecosystem modeling has significantly advanced our comprehension of the impacts of acidification on marine food webs. Machine learning algorithms enable the simulation of intricate interactions among diverse species and trophic levels, ranging from phytoplankton to apex predators. For instance, reinforcement learning models have been utilized to predict the behavioral and distributional adaptations of species in response to evolving ocean chemistry.

The effects on calcifying organisms, such as corals, mollusks, and certain plankton species, are of particular concern. Advanced computer vision and image processing algorithms analyze microscopic alterations in shell formation and skeletal structures, enabling the quantification of the impacts of reduced pH on these pivotal marine organisms. This data is subsequently integrated into broader ecosystem models, which predict potential tipping points and cascading effects.

The integration of AI with climate models has significantly advanced our understanding of feedback loops between ocean acidification and climate change. These models elucidate how alterations in ocean chemistry impact the ocean's capacity to absorb carbon dioxide, potentially accelerating both acidification and global warming. This comprehension is paramount for the development of effective mitigation strategies and policies.

Quantum computing, a relatively novel technology in scientific analysis and understanding, presents a transformative approach to enhancing current AI and ML capabilities. Its unique ability to process information based on quantum mechanical principles sets it apart. The fundamental advantage lies in quantum bits or qubits, which can exist in multiple states simultaneously through superposition, dramatically expanding computational possibilities compared to classical binary systems.

Quantum computing presents substantial advantages for managing high-dimensional data spaces. Traditional neural networks frequently encounter exponential computational demands as the number of features or parameters increase. Quantum neural networks, implemented through quantum circuits, can process this high-dimensional data more efficiently by exploiting quantum parallelism. For instance, quantum circuits can execute matrix operations fundamental to deep learning algorithms exponentially faster than classical computers in specific scenarios.

The concept of quantum entanglement plays a pivotal role in augmenting machine learning capabilities. Through entanglement, qubits can be correlated in ways lacking a classical counterpart, enabling quantum algorithms to discern and exploit patterns in data that may be imperceptible to classical ML approaches. This is particularly advantageous in analyzing intricate ocean ecosystem data, where subtle interactions among various factors can significantly influence acidification patterns.

Quantum variational algorithms represent a promising hybrid approach that combines classical and quantum processing. These algorithms employ quantum circuits as subroutines within larger classical optimization problems, making them particularly suitable for near-term quantum devices. In the context of ocean acidification modeling, these algorithms possess the potential to optimize intricate simulation parameters more efficiently than classical methods alone.

Quantum kernel methods extend classical machine learning kernels into the quantum domain. By mapping classical data into quantum feature spaces, these methods can capture more intricate relationships in the data. This capability holds particular significance for comprehending nonlinear interactions in ocean chemistry and biological systems, potentially unveiling previously unrecognized patterns in acidification processes.

Quantum-inspired algorithms, while executing on classical hardware, draw upon concepts from quantum computing to enhance performance. These algorithms have demonstrated promise in optimization problems and could augment the predictive capabilities of current machine learning models, enabling them to discern acidification trends and ecosystem responses more accurately. They serve as a bridge between current capabilities and the realization of full quantum implementation.

A notable advancement lies in quantum reservoir computing, which utilizes quantum systems as reservoirs for processing temporal data. This approach can potentially revolutionize time series analysis of oceanographic data, providing deeper insights into long-term acidification trends and their ecological impacts. The quantum reservoir's inherent evolution enables it to process complex temporal patterns more efficiently than classical recurrent neural networks.

Quantum computing with classical ML frameworks offers interesting possibilities. Hybrid quantum-classical architectures leverage quantum processors for computationally intensive tasks while maintaining classical reliability and accessibility. This approach is valuable for processing vast oceanographic data for acidification modeling.

Quantum machine learning enables real-time processing of complex ecosystem data, leading to more dynamic conservation strategies. Simulating quantum systems improves our understanding of molecular-level processes in marine organisms, providing insights into adaptation mechanisms and mitigation strategies.

Quantum-specific algorithms optimized for ocean science applications represent an emerging frontier. These algorithms better handle quantum processes involved in ocean acidification, leading to more accurate models and predictions. This includes better modeling of carbonate chemistry dynamics and interactions with biological systems at multiple scales.

The practical implementation of quantum computing in ocean acidification research necessitates sustained advancements in both hardware and software. As quantum computers attain greater power and reliability, their integration with existing machine learning frameworks is anticipated to accelerate, thereby providing increasingly sophisticated tools for comprehending and addressing this critical environmental challenge.

In the future, quantum computing and advanced AI algorithms hold the potential for even more sophisticated modeling capabilities, enabling us to simulate entire ocean ecosystems with unprecedented detail and accuracy. This advancement could be pivotal for developing targeted interventions to safeguard vulnerable species and maintain essential ecosystem services that humans rely upon.

Transparency is paramount for implementing evidence-based conservation measures and adapting coastal management strategies to protect marine resources and the communities that depend on them.

For humans, the implications of ocean acidification are profound and multifaceted. Machine learning models assist in forecasting the impacts on commercial fisheries by analyzing how acidification affects the larval development and survival rates of economically significant species. These models also predict alterations in coral reef systems, which are crucial for coastal protection and tourism. Natural language processing techniques are employed to analyze extensive scientific literature and research papers, thereby synthesizing our comprehension of acidification's broader socioeconomic effects.

CHAPTER TWO

Climate Science AI: Advance Concepts

Carbon Cycle Modeling

Carbon Sink Analysis and Emission Prediction

This domain of climate science has undergone numerous iterations, transitioning from conventional methods to the most recent machine learning techniques in tandem with the advancement of data collection technologies.

Conventional data collection commences with direct measurement systems, which serve as the foundation of traditional emission monitoring approaches. Ground-based monitoring stations are pivotal data collection points, equipped with sophisticated gas analyzers that continuously monitor concentrations of CO_2, CH_4, and other greenhouse gases. These stations are strategically positioned to capture both urban and rural atmospheric compositions. Eddy covariance towers represent another crucial component of direct measurement infrastructure, providing detailed measurements of gas exchange between ecosystems and the atmosphere. These towers utilize high-frequency wind measurements in conjunction with gas concentration data to calculate net ecosystem exchange.

A fundamental aspect of comprehending carbon sequestration dynamics commences with soil sampling and analysis. This process entails the collection of soil samples from various depths and landscapes, followed by laboratory analysis to ascertain organic carbon content, bulk density, and other pertinent parameters. These measurements are of paramount importance in calculating soil carbon stocks and monitoring temporal variations.

Forest inventories, in conjunction with applied biomass measurements, provide crucial data regarding above-ground carbon storage. Detailed field surveys are necessary, wherein researchers measure tree diameter, height, and species composition across representative plots. These measurements are subsequently converted to biomass estimates utilizing allometric equations. Vegetation sampling extends beyond trees to encompass understory vegetation, thereby providing a comprehensive view of ecosystem carbon storage.

Remote sensing technologies have revolutionized the scale and scope of emission and carbon sink monitoring. Satellite-based measurements utilizing instruments, such as MODIS, Landsat, and Sentinel, provide continuous, global-scale observations of land use, vegetation indices, and atmospheric composition. These platforms offer diverse spatial and temporal resolutions, enabling researchers to monitor both rapid changes and long-term trends in ecosystem properties and atmospheric conditions.

Spectral imaging technologies provide detailed insights into vegetation health and photosynthetic activity. By capturing reflected light across hundreds of narrow spectral bands, these systems can detect subtle variations in plant biochemistry and stress responses, providing early indicators of changes in carbon uptake capacity.

LiDAR (Light Detection and Ranging) surveys have emerged as a potent tool for three-dimensional mapping of vegetation structure. This technology provides precise measurements of canopy height, density, and structure, enabling accurate estimation of above-ground biomass across vast areas. The capability to penetrate canopy layers makes LiDAR particularly valuable for monitoring forest carbon stocks.

Economic and industrial data encompass a comprehensive spectrum of human activities that contribute to carbon emissions. Industrial facilities and power plants directly measure fossil fuel usage and emissions, including the type of fuel utilized and consumption rates. By employing these parameters, CO_2 emissions can be directly calculated.

In addition to traditional optical sensing, sophisticated methods for environmental monitoring have emerged. High-resolution satellite constellations provide frequent coverage of the Earth's surface, enabling near-real-time monitoring of environmental changes. Synthetic Aperture Radar (SAR) imaging offers all-weather monitoring capabilities, particularly valuable for regions frequently covered by clouds.

Multi-angle imaging spectroradiometers capture detailed information regarding atmospheric aerosols and their interactions with radiation. These instruments are pivotal for comprehending climate-forcing mechanisms. Extremely high-resolution radiometer systems facilitate continuous monitoring of surface temperature patterns and vegetation dynamics on global scales.

The integration of the Internet of Things (IoT) through sensor network technology has significantly enhanced environmental monitoring capabilities. Distributed sensor networks now enable real-time monitoring of greenhouse gas concentrations, meteorological conditions, and other environmental parameters across urban and natural landscapes. These networks typically comprise interconnected nodes equipped with diverse sensors, generating high-temporal resolution data streams that can be accessed and analyzed in real-time.

Industrial IoT devices and smart meters provide comprehensive energy consumption data at both individual and facility levels. These systems enable continuous monitoring of electricity usage patterns, facilitating the identification of opportunities for emission reduction and efficiency enhancements. Wireless sensor networks deployed in forest and agricultural areas collect data on soil moisture, temperature, and gas fluxes, providing insights into ecosystem responses to environmental changes.

By harnessing the comprehensive datasets amassed through conventional and contemporary methodologies, a robust framework for in-depth analysis is established. The cornerstone of this approach lies in statistical analysis, which forms the basis for interpreting emission and carbon sink data. Time series analysis techniques empower researchers to discern trends, seasonal variations, and cyclical patterns in emission data across diverse temporal scales. These methodologies encompass a range of approaches, including decomposition, smoothing, and forecasting, enabling the comprehension of both short-term fluctuations and long-term trends.

Regression analysis tools serve as pivotal instruments for identifying correlations between multiple variables influencing carbon emissions and sequestration. This encompasses both simple linear regression for fundamental relationship elucidation and multiple regression for intricate multi-variable analysis. Principal Component Analysis (PCA) assists researchers in reducing the dimensionality of substantial datasets while preserving essential information, facilitating the identification of key drivers of emission patterns.

The first method group comprises inventory-based approaches. These standardized frameworks quantitatively assess emissions and carbon sinks. The Intergovernmental Panel on Climate Change (IPCC) guidelines provide internationally recognized methodologies for calculating national greenhouse gas inventories. These methods encompass sector-specific calculation methodologies, emission factors, and uncertainty assessment procedures.

Life cycle assessment methods evaluate emissions across the entire product or process life cycle, from raw material extraction to end-of-life disposal. These assessments identify opportunities for emission reduction throughout supply chains and production processes. Carbon footprint calculations extend this approach to organizations, products, and activities, providing standardized metrics for comparing emission impacts.

Physical modeling tools are an effective approach to comprehending the mechanistic processes governing carbon cycles and emission pathways. Atmospheric transport models simulate the movement and dispersion of greenhouse gases through the atmosphere, accounting for factors such as wind patterns, temperature gradients, and chemical reactions. These models are crucial for comprehending the spatial distribution of emissions and their atmospheric fate.

Biogeochemical cycle models integrate various physical, chemical, and biological processes to simulate carbon exchange between distinct environmental compartments. These models incorporate factors, such as photosynthesis, respiration, decomposition, and nutrient cycling to predict ecosystem carbon dynamics. Ocean carbon cycle models specifically focus on marine systems, simulating processes such as air-sea gas exchange, biological pumping, and ocean circulation patterns that influence carbon sequestration.

We must incorporate AI and machine learning techniques to further elucidate the impact on climate. This does not imply that the previously described techniques are devoid of machine learning algorithms. The terms "ML" and "AI" merely serve to clarify specific tools primarily derived from more recent advancements and labeled as learning methods.

Supervised learning employs labeled data, wherein the correct outputs are predetermined during training (similar to teaching with pre-assigned answers). Conversely, unsupervised learning operates with unlabeled data to independently identify latent patterns and structures (analogous to discovering groupings without prior knowledge).

These constitute fundamental methodologies employed by machine learning algorithms in acquiring knowledge from data.

The initial section delves into regression methodologies. Random Forest Regression stands out for its ability to handle intricate, non-linear relationships in environmental data. It effectively incorporates multiple predictive variables while maintaining robustness against outliers. Support Vector Regression provides robust tools for trend analysis, particularly effective in managing high-dimensional datasets prevalent in environmental monitoring.

Gradient Boosting Machines provide sophisticated pattern recognition capabilities, systematically enhancing predictions by prioritizing previously poorly-predicted instances. Deep learning architectures excel at capturing intricate temporal and spatial patterns in emission data, thereby enabling more accurate long-term forecasting capabilities.

Satellite image analysis by CNNs enables automated detection of land use changes, industrial facilities, and ecosystem characteristics. These networks can process vast amounts of spatial data to identify patterns and changes that human analysts might miss.

Classification algorithms play pivotal roles in categorizing emission sources and land use patterns. Decision trees provide interpretable classification frameworks, which are particularly useful for identifying emission source categories based on multiple parameters. Support Vector Machines excel at land use classification tasks, effectively distinguishing between different ecosystem types using remote sensing data.

The unsupervised learning approaches unveiled hidden patterns in emission and carbon sink data. Clustering algorithms facilitated the identification of natural groupings within emission patterns, potentially revealing previously unknown connections between various sources or regions. Dimensionality reduction techniques effectively managed the complexity of substantial environmental datasets while preserving crucial information for analysis.

Anomaly detection methods identified anomalous patterns in emission data, enabling the detection of potential measurement errors or unexpected alterations in emission sources. These techniques are particularly valuable for quality control within extensive monitoring networks and for identifying emerging trends or issues necessitating further investigation.

Contemporary deep learning methodologies have substantially enhanced emission forecasting capabilities. LSTM excels at capturing long-term dependencies in emission patterns, thereby enabling more precise predictions of future trends. Attention mechanisms facilitate the identification of pivotal factors influencing emission patterns, while Transformer models capture intricate temporal relationships in environmental data.

The preprocessing stage is paramount in any scientific discipline that involves analyzing collected data. The methods employed directly depend on the consolidated data for a reliable prediction. A comprehensive suite of quality control procedures and missing data imputation techniques ensure the continuity of data records, while outlier detection methods identify and manage anomalous measurements that could potentially distort analysis results.

Data standardization and normalization procedures facilitate interoperability among disparate measurement sources and temporal scales. Feature engineering processes construct pertinent derived variables that encapsulate crucial aspects of emission and sequestration processes, thereby enhancing subsequent analysis stages.

Model development adheres to systematic methodologies to guarantee reliability and reproducibility. Cross-validation techniques facilitate the assessment of model performance across diverse conditions and time periods, while hyperparameter optimization ensures optimal model configuration for specific applications. Ensemble methods amalgamate multiple modeling approaches to augment prediction accuracy and reliability.

Uncertainty quantification provides crucial information regarding the confidence and reliability of predictions. It encompasses both aleatory uncertainty arising from natural variability and epistemic uncertainty attributable to model limitations and data quality concerns.

Methodological considerations include careful selection of appropriate analysis approaches for specific applications. This involves balancing model complexity with data availability and quality, considering scale effects in both measurements and modeling, and incorporating domain knowledge into analysis frameworks.

Uncertainty propagation through analysis pipelines necessitates meticulous attention to ensure the reliability of results. This encompasses the consideration of measurement uncertainties, model limitations, and the impact of various assumptions and simplifications employed in analysis procedures.

Upon the successful completion of these meticulously described steps, the validation of the analysis commences. This involves comparing the results with ground truth measurements, conducting cross-validation against independent data sources, and verifying the results against established physical principles and constraints. Model validation procedures assess the performance of the model across diverse conditions and scales, thereby ensuring its reliable application in various situations.

Technical challenges, such as addressing the computational requirements for processing large-scale environmental data by running complex models, are among the factors contributing to these tasks. Data storage and management must consider diverse data types without compromising accessibility and security.

Industry Production Emission Modeling

Emission modeling for industrial production relies heavily on sensor data, including real-time measurements of greenhouse gases, particulate matter, and other pollutants emitted from industrial facilities. This data is complemented by operational parameters such as temperature, pressure, fuel consumption, and production rates. Historical weather data and atmospheric conditions are also incorporated to comprehend the environmental impacts.

Previously, emission modeling relied on statistical methods and fundamental regression analysis. Gaussian plume models were prevalent for dispersion calculations, while box models offered simplified representations of atmospheric chemistry. These approaches frequently encountered challenges in complex industrial processes and necessitated extensive manual calibration.

AI approaches predominantly utilize deep learning models, particularly recurrent neural networks (RNNs) and long short-term memory (LSTM) networks, which excel in processing time-series data. These models possess the capability to predict emission patterns and identify anomalies in real-time. Transformer architectures are increasingly gaining traction due to their ability to handle long-term dependencies within emission data sequences.

In contemporary industrial emission modeling, machine learning techniques have undergone substantial advancements. Deep learning architectures like LSTMs and Transformers process intricate temporal patterns in emission data. These models simultaneously analyze multiple data streams, including real-time sensor readings, operational parameters, and environmental conditions.

CNNs are used to recognize spatial emission patterns, particularly in facilities with multiple emission sources. They facilitate the identification of emission hotspots and the prediction of dispersion patterns. Transfer learning techniques enable models trained on data from one facility to be adapted for others with similar processes, addressing the common challenge of limited historical data.

Reinforcement learning algorithms optimize production processes in real-time, striking a balance between emission reduction and production efficiency. These systems learn from historical control decisions and their outcomes, continuously refining their recommendations for process parameters.

Ensemble methods combine multiple models to enhance prediction accuracy and robustness. Random forests and gradient boosting machines collaborate with deep learning models, each capturing distinct aspects of emission patterns. This hybrid approach enables handling both linear and non-linear relationships in emission data.

Unsupervised learning techniques, particularly autoencoders and clustering algorithms, detect anomalous emission patterns and identify underlying process states contributing to elevated emissions. These insights empower operators to proactively prevent potential environmental compliance issues.

Federated learning emerges as a solution for sharing emission modeling insights across facilities while preserving data privacy, thereby facilitating industry-wide improvements in emission reduction strategies.

Recent advancements include attention mechanisms for identifying critical factors in emission formation and graph neural networks for modeling intricate interactions between various production units and their collective impact on emissions.

The transition to AI-driven approaches has significantly enhanced accuracy and enabled predictive capabilities.

Modern systems can forecast emission levels, optimize production processes for environmental impact, and proactively identify potential violations.

This evolution has been particularly pivotal for industries such as steel manufacturing, cement production, and power generation, where emissions modeling is paramount for regulatory compliance and environmental management.

Furthermore, the integration of machine learning has enhanced the management of uncertainty and variability in industrial processes, transcending the deterministic approaches of traditional models to provide probabilistic forecasts and risk assessments.

Forest Cover Change Detection

Earth observation data from satellite imagery revealed a forest loss rate of 2,101 km2 per year. Notably, deforestation in Brazil has been significantly reduced, as extensively documented. However, this reduction was offset by an increase in forest loss in Indonesia. Despite the availability of data, we still lack comprehensive information on spatially and temporally explicit patterns at a global scale. Based on the available data, the tropics were the only domain exhibiting a statistically significant annual forest loss trend.

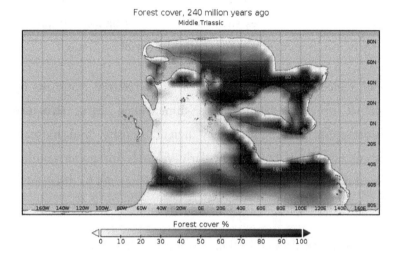

Forest cover, 240 million years ago
Middle Triassic

Forest cover %

The northwestern United States is an area of intensive forestry, as is the entire range of temperate Canada. The intermountain west of North America exhibits a loss dynamic primarily attributed to fire, logging, and disease. For instance, large-scale tree mortality due to mountain pine beetle infestation, most evident in British Columbia, Canada, has contributed to this loss. [5]

Measuring and predicting trends in forest cover necessitates the utilization of a diverse array of data sources and analytical methodologies that have undergone substantial evolution over time. The primary data sources encompass satellite imagery from various platforms, such as Landsat, Sentinel, and MODIS. These

---

[5] Numbers are from a publication "Observing the forest and the trees, 13.August 2013."

satellites provide multispectral and hyperspectral imagery at diverse spatial and temporal resolutions. These satellites capture visible light and infrared bands, which are crucial for vegetation analysis through indices such as NDVI (Normalized Difference Vegetation Index) and EVI (Enhanced Vegetation Index).

Data collected through both aerial and terrestrial platforms by LiDAR provides detailed three-dimensional forest structure information, including canopy height, density, and biomass estimates. This data is particularly valuable for understanding forest degradation and subtle changes that may not be discernible in traditional satellite imagery.

Ground-based data collection methods, such as forest inventory plots, species composition surveys, and detailed measurements of individual trees, are crucial for understanding environmental factors influencing forest health and growth. Weather station data, including precipitation, temperature, and humidity measurements, provides valuable insights into these factors. Soil samples and analysis offer insights into forest ecosystem health and potential regeneration.

Traditionally, forest cover analysis relied heavily on the manual interpretation of aerial photographs and basic image processing techniques. These methods included supervised and unsupervised classification of satellite imagery using statistical approaches such as maximum likelihood classification. Change detection was performed through simple band arithmetic and post-classification comparison techniques.

The advent of machine learning has revolutionized analysis capabilities, particularly in the realm of forest monitoring. Support Vector Machines (SVM) and Random Forests have emerged as prominent algorithms due to their ability to handle intricate, high-dimensional data and deliver robust classification outcomes. These algorithms excel at distinguishing various forest types and detecting alterations in forest cover by drawing upon labeled training data that encapsulates known forest conditions.

In contrast, contemporary deep learning approaches have catalyzed a paradigm shift in forest monitoring. CNNs are employed to identify deforestation, forest degradation, or regeneration. Initially designed for medical image segmentation, U-Net architectures have been repurposed for precise forest cover mapping and change detection.

Time series analysis enhanced by RNNs and LSTM networks is used to detect gradual changes and predict future forest cover trends. These models capture seasonal variations and distinguish them from permanent changes.

Transfer learning adapts models trained on well-documented forest regions to areas with limited data, improving global forest monitoring. Ensemble methods combine multiple models, integrating traditional machine learning with deep learning approaches, to enhance prediction accuracy and robustness.

AI for forest monitoring has significantly enhanced the ability to identify and analyze relevant spatial and temporal patterns. Graph neural networks, in particular, have proven instrumental in modeling complex interactions between diverse forest patches and environmental factors. These sophisticated techniques are particularly valuable in understanding forest fragmentation and connectivity.

Cloud computing platforms and advanced big data technologies have enabled the processing and analysis of vast amounts of forest-related data on a global scale. Google Earth Engine, for instance, has democratized access to satellite imagery and processing capabilities, facilitating near-real-time forest monitoring across extensive geographic regions.

Modern systems also integrate multiple data sources through data fusion techniques. For instance, combining optical satellite imagery with radar data (Synthetic Aperture Radar, or SAR) enables continuous monitoring even in cloudy conditions, which is particularly crucial in tropical regions characterized by frequent cloud cover.

Predictive modeling of forest cover changes increasingly incorporates socioeconomic data, including population density, economic indicators, and land use policies. This integration facilitates the understanding and prediction of human-induced changes in forest cover, enabling more effective conservation strategies and policy interventions.

The accuracy of forest cover prediction has significantly improved with technological advancements. However, challenges persist, including the need for improved methods to handle data uncertainty, enhanced techniques for detecting subtle forms of forest degradation, and more efficient data processing methods due to the increasing volume of available data. Ongoing research focuses on developing more sophisticated AI models that can capture the complex dynamics of forest ecosystems while maintaining computational efficiency and interpretability for forest managers and policymakers.

Future research priorities address current limitations and expand analysis capabilities. This includes improving uncertainty quantification methods, enhancing spatial and temporal resolution of analyses, and developing better approaches for integrating multiple data sources. Advanced model interpretability techniques will enhance understanding of model behavior and reliability, while real-time analysis capabilities will enable more responsive monitoring and management systems.

Global Impact of Climate Change

Agricultural Yield Assessment and Prediction

Accurately assessing and predicting agricultural yields is paramount for ensuring food security, maintaining economic stability, and optimizing agricultural management practices. Historically, yield estimation relied on conventional methods such as farmer surveys, field sampling, and statistical models derived from historical data. However, the escalating complexity of agricultural systems and the escalating impacts of climate change necessitate more sophisticated approaches. Consequently, the integration of advanced data collection techniques and sophisticated ML and AI methodologies into agricultural yield assessment and prediction has become imperative.

To ensure sustainable and efficient utilization of natural research, it is crucial to select the parameters to assess the environmental conditions that serve as the ground truth for developing reliable and transparent prediction models. Weather parameters are essential, as are agronomic principles of a plant, for a final estimation. Like statistical models, machine learning algorithms can utilize the outputs of other methods as features. Furthermore, machine learning algorithms possess distinct advantages, such as the ability to model non-linear relationships between multiple data sources.

Large-scale crop yield forecasting systems, such as the MARS Crop Yield Forecasting System (MCYFS) of the European Commission's Joint Research Centre (JRC) and the National Agricultural Statistics Service (NASS) of the United States Department of Agriculture (USDA), possess the necessary infrastructure and historical data to construct and evaluate crop yield prediction models for various crops and locations. However, the operational systems currently in use do not incorporate machine learning algorithms. Instead, they construct statistical models based on weather observations, field survey results, crop growth model outputs, remote sensing indicators, and yield statistics.

Prior to exploring contemporary techniques, it is imperative to recognize the foundational methods employed for decades. This entails directly obtaining information from farmers regarding their anticipated or realized yields. While this method provides valuable local knowledge, it can be subjective and susceptible to biases.

Another approach involves physically collecting samples from fields and measuring yield components (such as the number of plants and grain weight). This method offers high accuracy but is labor-intensive and limited in terms of spatial coverage.

As a consistent practice, historical yield data and environmental factors (such as weather) are utilized to establish statistical relationships and forecast future yields. These models prove useful for broad-scale predictions, but they may not adequately capture local variations or the effects of novel technologies.

These methods were further enhanced by incorporating novel technologies for data collection in agriculture, thereby enabling the acquisition of more comprehensive, precise, and timely information. Satellite imagery, aerial photography (including drones), and other remote platforms are employed to gather data on crop growth, health, and environmental conditions. Plant health assessment, biomass estimation, and stress impact analysis are accomplished through hyperspectral and multispectral imaging. Thermal imaging is utilized to monitor canopy temperature and detect water stress and other physiological conditions.

3D models of fields and crops, incorporating information on plant height, biomass, and density, are generated through laser scanning.

Climate impact on Jayatataka Baray

Soil sensors specialize in measuring soil moisture, temperature, electrical conductivity, and nutrient levels. Plant sensors, on the other hand, conduct measurements to monitor plant growth, leaf area chlorophyll content, and other physical attributes. Weather stations provide comprehensive data on temperature, humidity, rainfall, wind speed, and solar radiation. All of these sensors are ground-based.

Climate change poses substantial challenges to agricultural production, characterized by altering temperature and precipitation patterns, escalating the frequency of extreme weather events, and rising CO_2 levels. ML and AI can play a pivotal role in assessing and mitigating these impacts.

To effectively utilize the gathered data, establish dependencies, derive insightful conclusions, and subsequently make precise predictions, it is imperative to employ machine learning and artificial intelligence tools.

Regression models can predict yield in accordance with relationships established with management and environmental factors. Decision trees and random forest algorithms are employed to identify intricate, non-linear relationships and crucial predictors.

Support vector machines can classify effectively and predict yield based on various features. Pattern neural networks can be used to model large datasets with highly complex interactions.

A subfield of machine learning that utilizes artificial neural networks with multiple layers to learn intricate patterns from extremely large datasets, particularly useful for analyzing image data from remote sensing.

The integration of advanced data collection techniques and sophisticated machine learning and artificial intelligence methods is revolutionizing agricultural yield assessment and prediction. These technologies provide more precise, timely, and spatially detailed information, empowering farmers, policymakers, and other stakeholders to make informed decisions. By incorporating climate change considerations into these approaches, we can develop more resilient and sustainable agricultural systems that can effectively address the escalating global food demand amidst climate change. Ongoing advancements in these domains will further enhance our capacity to comprehend and manage the intricate interactions within agricultural systems, ensuring food security for the future.

Population Displacement: Projection

One of the most intricate and unfortunate consequences of global climate change for humanity is the phenomenon of population displacement. We present a likely reliable projection for the next 25 years by analyzing climate-driven migration patterns and current trends.

Numerous interconnected events are the primary drivers of population displacement. The most immediate and evident factor is the escalating frequency and intensity of extreme weather events. Hurricanes, floods, and tropical cyclones can render entire regions uninhabitable within hours or days, necessitating immediate evacuation and often resulting in permanent displacement when the damage exceeds the capacity of communities to rebuild. The aftermath of Hurricane Maria in Puerto Rico exemplifies this pattern, where the destruction of critical infrastructure compelled approximately 130,000 residents to relocate to the mainland United States.

Subtle alterations in the environmental degradation process lead to displacement. These alterations include rising sea levels, which gradually erode coastlines and contaminate freshwater aquifers with saltwater intrusion, rendering coastal areas increasingly uninhabitable. Low-lying island nations in the Pacific, such as Kiribati and Tuvalu, are already experiencing existential threats due to this process. Agricultural regions face mounting pressure from changing precipitation patterns and increasing temperatures, which can result in crop failures and soil degradation. The expansion of arid zones in sub-Saharan Africa has already commenced, displacing farming communities to urban areas or more fertile regions.

In developing countries, the combination of heat island effects and inadequate infrastructure makes cities increasingly vulnerable to extreme heat events. When combined with water scarcity and stressed power grids, these conditions can render dense urban areas uninhabitable during peak summer months, forcing seasonal or permanent migration.

Socioeconomic and political pressures, coupled with economic disparities, often determine who can adapt in place and who must relocate. Vulnerable populations are disproportionately affected by displacement.

These interconnected forces, such as resource competition, do not occur in isolation and potentially trigger more severe movements, including secondary displacements.

Rural communities are often the initial recipients of environmental degradation, leading to migration towards urban centers. However, these cities frequently lack the necessary infrastructure and resources to accommodate the rapid influx of populations, resulting in the formation of informal settlements that are susceptible to their own set of environmental challenges. This dynamic between urban and rural areas significantly influences displacement patterns.

The urban-rural dynamic plays a pivotal role in shaping displacement patterns. This phenomenon has been observed in Bangladesh, where rural coastal communities displaced by flooding and saltwater intrusion often find themselves in precarious urban environments.

The methodology for creating climate-related displacement projections involves the integration of various models and data sources. This process entails a review of existing tools and, as necessary, their extension.

Initially, displacement projections commence with climate models, particularly those from the Coupled Model Intercomparison Project Phase 6 (CMIP6).[6] These models provide projections of pivotal variables such as temperature, precipitation, and sea level rise under diverse emission scenarios. The Shared Socioeconomic Pathways (SSPs) framework offers narratives of

---

[6] Working Group on Coupled Modeling (WGCM) under World Climate Research Programme.

future socioeconomic development that can be harmonized with climate scenarios to construct more comprehensive projections.

GIS enables the identification of vulnerable regions by exploring potential migration corridors. Utilizing elevation, population density, infrastructure, and projected climate impacts, researchers can discern both areas of high displacement risk and potential destination regions. The Environmental Systems Research Institute (ESRI) offers specialized tools for climate vulnerability mapping that can be seamlessly integrated with population data.

A reliable projection of climate displacement necessitates the integration of disparate data sources. Historical displacement data from the Internal Displacement Monitoring Centre (IDMC) serves as a baseline, elucidating displacement patterns and their correlation with environmental events. Satellite imagery and remote sensing data enable real-time monitoring of environmental transformations and human settlement patterns. Census data and demographic surveys provide insights into population distribution and socioeconomic factors that predispose individuals to displacement vulnerability.

Advanced machine learning tools have facilitated more intricate analysis of historical displacement patterns and their correlation with environmental factors. Neural networks can be trained on historical data to discern intricate relationships between environmental alterations and population movements. However, it is imperative to account for the unprecedented nature of future climate change.

Human decisions are intricate and do not necessarily indicate a specific migration pattern. Stress levels have a flexible impact, with some individuals attempting to adopt while others may relocate due to limited resources. The interconnected nature of climate impacts makes it challenging to isolate climate-specific displacement from other forms of migration. Furthermore, tipping points in both environmental and social systems can lead to non-linear changes that are difficult to predict using current models.

Integrated assessment models that integrate climate, environmental, and socioeconomic factors at multiple scales must be developed. Investments in enhanced data collection systems, particularly in vulnerable regions, are necessary to better comprehend displacement patterns and triggers. Furthermore, early warning systems should be strengthened to facilitate anticipation and preparedness for displacement events. Additionally, local knowledge and community-level adaptation strategies need to be incorporated into projection models.

It is infeasible to achieve a complete and reliable projection of climate-induced displacement due to the potential for significant unforeseen impacts arising from independent events that may not be apparent at the outset.

By integrating advanced modeling techniques, enhanced data collection methods, and comprehensive analysis approaches, we can effectively anticipate and prepare for future displacement patterns. This comprehension is paramount for developing effective adaptation strategies and supporting systems for impacted communities.

Disease Spread Modeling

The interdependence of climate change and disease transmission poses a complex challenge for global public health. As our planet experiences shifting temperature patterns, altered precipitation cycles, and increasingly frequent extreme weather events, the landscape of infectious and non-infectious diseases undergoes significant transformation. This analysis examines current trends and projected changes in disease patterns as influenced by climate change.

Climate change-induced air quality degradation will have substantial impacts on respiratory health through various mechanisms. Rising temperatures and altered precipitation patterns alter the timing and duration of pollen seasons. Consequently, many regions are witnessing extended and more severe allergy seasons. Some research indicates that pollen production could potentially increase by 200% in certain areas. This trend will disproportionately affect individuals with asthma and other respiratory conditions.

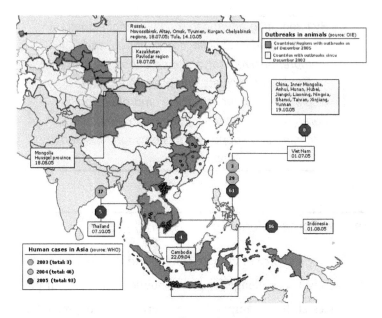

Avian Influenza Spread

Ground-level ozone, a crucial component of smog, is more readily formed at elevated temperatures. As climate change brings more frequent heatwaves, urban areas are particularly susceptible to increased risks of high ozone days, resulting in elevated rates of respiratory distress and cardiovascular complications.

Warmer temperatures and drought conditions contribute to more frequent and intense wildfires, leading to increased exposure to particulate matter and other air pollutants. The health impacts extend far beyond the immediate fire zones, as smoke can travel thousands of miles. These conditions exacerbate existing respiratory diseases and can lead to new cases of chronic respiratory illness.

As vector-borne diseases such as mosquito-borne illnesses spread, dengue fever is expanding into previously unexplored territories. As temperatures rise, regions previously too frigid for Aedes aegypti and Aedes albopictus mosquitoes become suitable habitats. Models indicate that by 2050, an additional two billion people could reside in areas where these diseases are endemic. Furthermore, the seasonal transmission window for these vectors is expanding in many regions, enabling extended periods of disease transmission.

Historically confined to specific geographic zones, malaria is now emerging at higher altitudes in regions such as the East African highlands. The Anopheles mosquito's range is expanding due to rising temperatures, which render previously inhospitable areas suitable for their survival. Furthermore, the parasites responsible for malaria develop more rapidly within mosquitoes at higher temperatures, potentially elevating transmission rates within existing malaria zones.

Tick-borne diseases are expanding their reach due to climate change. Lyme disease, which is carried by Ixodes ticks, is now popping up in parts of Canada and Northern Europe that used to be tick-free. The ticks are more active now and are spreading northward. This is happening with other tick-borne illnesses like tick-borne encephalitis and Rocky Mountain spotted fever too.

Climate change is changing our water supply and causing some serious health problems. More extreme rainfall events followed by droughts create perfect conditions for cholera outbreaks. The

Vibrio bacteria that cause cholera love warm water, and flooding can contaminate drinking water, spreading the disease like wildfire. Coastal areas are especially at risk because rising sea levels are mixing freshwater with saltwater, making it a breeding ground for Vibrio species.

Schistosomiasis, a parasitic disease spread through freshwater snails, is also rising. Warming temperatures are making new areas suitable for the intermediate host snails, and changes in farming and water management due to climate change could create new habitats for these disease vectors.

Harmful algal blooms, while not diseases themselves, are becoming more common and intense because of warmer waters and more nutrients from extreme rainfall. These blooms can produce toxins that cause various illnesses and contaminate drinking water, like what happened in Toledo, Ohio's water crisis. We can expect these events to get even more frequent and severe in the coming decades.

Infectious diseases are reemerging due to climate change, creating favorable conditions. This trend will continue and potentially increase the risk of new diseases. Permafrost thawing in Arctic regions may release ancient microorganisms that modern immune systems have never encountered. While the full implications are not yet understood, this process represents a novel disease risk that requires careful monitoring.

Fungal diseases are already exhibiting concerning patterns of adaptation to warmer temperatures. Cryptococcus gattii, traditionally found in tropical and subtropical regions, has emerged in temperate areas such as the Pacific Northwest. Some researchers suggest that rising temperatures could lead to the evolution of fungal pathogens that are more capable of surviving at human body temperature, potentially creating new threats to human health.

As climate change disrupts animal migration patterns and brings wildlife into closer proximity with human populations, the incidence of zoonotic diseases is expected to increase. Ecosystem disruptions force animals to seek new habitats, potentially creating novel disease transmission pathways.

These developments pose significant challenges for healthcare systems globally. Regions may encounter diseases for which they have limited medical expertise, necessitating rapid adaptations in medical training and resource allocation. The geographic expansion of various diseases necessitates healthcare systems to maintain readiness for a broader range of conditions than previously anticipated.

Infrastructure challenges include the need for enhanced disease surveillance systems, particularly in regions newly at risk for specific diseases. Water treatment facilities must be upgraded to address new threats, and urban planning must increasingly incorporate disease vector control into design and maintenance strategies.

The complex relationship between environmental changes and disease spread needs constant monitoring and flexible response plans.

While current predictions show possible trends, these systems are constantly changing, so public health planning needs to be flexible and prepared. To overcome these challenges, we need global cooperation and resource allocation to protect public health in a changing climate.

Economic Impact Prediction

The global economic impact of climate change is profound and intricate, as it is interwoven with physical, social, and financial factors. Machine learning algorithms and AI techniques have significantly enhanced our broader perspective and forecasting capabilities. Deep learning models analyze historical climate and economic data to identify patterns and correlations previously undetectable through conventional statistical methods. Neural networks process vast datasets of weather patterns, agricultural yields, energy consumption, and market responses to environmental changes. These models account for complex feedback loops between climate events and economic outcomes, providing more nuanced predictions than earlier forecasting methods.

Satellites, ground sensors, and IoT devices actively monitor environmental changes. These systems provide unprecedented precision in tracking sea level rise, temperature variations, precipitation patterns, and extreme weather events. Advanced remote sensing technologies, coupled with machine learning algorithms, enable real-time monitoring of glacial retreat, deforestation, and alterations in agricultural productivity. This data serves as the foundation for constructing intricate models that project future developments.

Coastal economies are in for a tough time. Rising sea levels and stronger storms are causing a lot of damage. Scientists are using complex models to predict how much infrastructure will be destroyed, how many people will have to move, and how much businesses will lose. They say that big coastal cities will need to spend a lot of money to protect themselves, and some low-lying areas might not be worth living in anymore. It will cost the government and private companies a lot of money to protect the coast.

The energy sector transformation presents another critical economic dimension. Machine learning algorithms optimize the integration of renewable energy into power grids, predict maintenance requirements for infrastructure, and model the economic viability of various energy sources under diverse climate scenarios. These models indicate accelerating cost advantages for renewable energy technologies, although the transition timeline varies substantially by region and economic capacity.

Agricultural impacts pose significant economic challenges. AI-powered crop yield prediction models incorporate soil moisture data, temperature trends, and precipitation forecasts to project food production capabilities. These models suggest that changing precipitation patterns and temperature extremes will reduce yields in many current agricultural regions while potentially opening new farming opportunities in previously unsuitable areas. This geographic shift in agricultural productivity will likely reshape global trade patterns and food security dynamics.

The economic implications of healthcare are becoming increasingly evident through the analysis of climate-health relationships using artificial intelligence. Models predict rising healthcare costs due to heat-related illnesses, alterations in disease patterns, and the effects of air pollution. These health consequences have substantial implications for labor productivity and public health expenditures.

Machine learning algorithms are employed to analyze global supply networks, identifying climate-sensitive points crucial for comprehending and addressing challenges within the supply chain. It is imperative to synchronize global and regional distribution workflows. By leveraging this knowledge, disruptions in the supply chain can be minimized, and catastrophic failures can be averted.

Climate change-induced migration patterns will reshape labor markets and urban development.

The aggregate economic impact over the next 30 years remains uncertain, despite advanced modeling capabilities. Current projections suggest global GDP losses between 10% and 23% compared to a scenario without climate change, with impacts varying significantly by region and adaptive capacity. Regions with lower adaptive capacity generally face greater relative impacts due to higher exposure to climate risks.

Machine learning models analyzing historical migration responses to environmental changes project substantial population movements, particularly from areas facing water stress or extreme heat. These demographic shifts will influence housing markets, public service demands, and regional economic development patterns.

The transition risks associated with the decarbonization of economies further complicate economic forecasting. Machine learning models analyzing policy scenarios, technological advancements, and market responses indicate substantial near-term adjustment costs but also potential long-term benefits from averted climate damages and the emergence of new economic opportunities in low-carbon sectors.

These projections underscore the paramount importance of adaptation investments and emission reduction efforts in determining future economic trajectories. While advanced measurement and modeling capabilities provide increasingly sophisticated insights into climate-economy interactions, they also reveal the magnitude of the challenge and the urgency of response measures.

The role of AI and ML in comprehending and addressing these challenges is likely to expand, with ongoing improvements in prediction accuracy and decision support capabilities. Nevertheless, these tools remain auxiliary to human decision-making rather than autonomous solutions, emphasizing the ongoing significance of policy choices and collective action in shaping economic outcomes.

The infrastructure necessitates adaptation, which poses a challenge in predicting whether the adjustment can be achieved linearly or if more significant alterations will be required. Given the interdependence of numerous variables with uncertain scale changes, a comprehensive projection for the next three decades and beyond remains virtually impossible.

Emerging Technologies and Methods

Advanced Sensor Networks

The evolution of advanced sensor networks represents a transformative force in our ability to monitor and predict climate change over the coming decades. These sophisticated networks, comprising interconnected arrays of high-precision instruments, are revolutionizing our understanding of climate dynamics by providing unprecedented granularity in environmental data collection. As we look toward 2050-2060, the deployment of these systems is expected to expand dramatically, creating an increasingly dense web of measurement points across terrestrial, marine, and atmospheric environments.

Current advanced sensor networks incorporate a diverse array of measurement capabilities, from basic meteorological parameters like temperature and humidity to complex atmospheric chemistry analyses. These systems are increasingly being enhanced with artificial intelligence and machine learning capabilities, allowing for real-time data validation and preliminary analysis at the point of collection. The next generation of sensors, currently in development, will feature even more sophisticated capabilities, including quantum-based measurements for unprecedented precision in detecting greenhouse gases and other climate-relevant parameters.

A particularly promising development in sensor network technology is the integration of bio-inspired sensors that mimic natural systems' ability to detect subtle environmental changes. These biomimetic sensors are expected to become increasingly prevalent by 2040, offering new ways to monitor ecosystem responses to climate change. For instance, sensors modeled after plant stomata are being developed to measure CO_2 flux more accurately than conventional methods, while artificial leaf systems can provide detailed information about local microclimate conditions.

The marine environment presents unique challenges and opportunities for advanced sensor networks. By 2050, we anticipate the deployment of vast networks of autonomous underwater vehicles equipped with advanced sensor arrays, capable of monitoring ocean acidification, temperature profiles, and marine ecosystem health across unprecedented spatial and temporal scales. These systems will be crucial for understanding ocean-atmosphere interactions and their role in global climate dynamics.

In urban environments, advanced sensor networks are expected to evolve into comprehensive environmental monitoring systems that integrate air quality, urban heat island effects, and local weather patterns. These networks will become increasingly important as cities adapt to climate change, providing crucial data for urban planning and emergency response systems. By 2060, urban sensor networks are predicted to achieve near-complete coverage of major metropolitan areas, offering real-time environmental monitoring at street-level resolution.

The integration of satellite-based remote sensing with ground-based sensor networks will create a multi-layered observation system capable of detecting and tracking climate change impacts across multiple scales. This fusion of data sources will be particularly valuable for monitoring critical regions such as the Arctic, where rapid environmental changes have global implications. Advanced sensor networks in these regions will provide early warning signals of significant climate system changes, such as permafrost thaw or sea ice dynamics.

Looking toward 2050, we can expect sensor networks to become increasingly autonomous and self-organizing, with the ability to adaptively modify their measurement strategies based on observed conditions. This flexibility will be crucial for capturing extreme weather events and other climate-related phenomena that require rapid response and detailed monitoring. The development of self-healing network architectures will ensure resilience in the face of increasingly severe weather conditions, maintaining data continuity even under challenging circumstances.

The miniaturization of sensor technology, coupled with advances in energy harvesting and storage, will enable the deployment of vast numbers of small, independent sensing nodes. These systems will form the backbone of a global environmental monitoring network that provides continuous, high-resolution data on climate-relevant parameters. By 2060, it's anticipated that these networks will achieve global coverage, including remote and previously undersampled regions, providing a complete picture of Earth's changing climate system.

Data management and integration will become increasingly sophisticated as sensor networks expand. Advanced artificial intelligence systems will be essential for processing the massive volumes of data generated by these networks, identifying patterns and trends that might otherwise go unnoticed. The development of edge computing capabilities within sensor networks will enable real-time data processing and analysis, reducing latency and improving the timeliness of climate-related alerts and predictions.

The social implications of advanced sensor networks in climate monitoring are significant. By 2050, public access to real-time environmental data is expected to become standardized, leading to increased awareness and engagement with climate issues at the community level. This democratization of climate data will support more informed decision-making and help drive adaptation strategies at local and regional scales.

Real-time Climate Analysis Systems

Real-time climate analysis systems represent a revolutionary advancement in our ability to understand and respond to climate change as it happens. These sophisticated platforms combine continuous data streams, advanced analytics, and high-performance computing to provide immediate insights into climate dynamics. As we look toward the 2050-2060 timeframe, these systems will become increasingly crucial for climate adaptation and mitigation strategies.

The foundation of modern real-time climate analysis lies in the integration of multiple data sources, including satellite observations, ground-based measurements, ocean buoys, and atmospheric sensors. These systems are evolving beyond simple data collection and visualization, incorporating advanced machine learning algorithms that can detect subtle patterns and anomalies in climate data as they emerge. By 2040, we expect these systems to achieve predictive capabilities that can anticipate climate-related events hours to weeks in advance with unprecedented accuracy.

A significant development in real-time climate analysis is the emergence of neural network architectures specifically designed for processing climate data streams. These specialized AI systems can handle the complexity and interdependencies of climate variables, identifying correlations and causations that might escape traditional analytical methods. The next generation of these systems, expected to be fully operational by 2045, will incorporate quantum-inspired algorithms that can process climate data with even greater sophistication, enabling the detection of complex climate patterns across multiple temporal and spatial scales.

The integration of real-time climate analysis with emergency response systems represents a crucial advancement in climate adaptation. By 2050, these integrated platforms are expected to provide automated early warning systems for extreme weather events, enabling communities to respond more effectively to climate-related emergencies. These systems will incorporate social and economic data alongside climate measurements, providing a more comprehensive understanding of climate impacts on human systems and infrastructure.

Urban environments present unique challenges and opportunities for real-time climate analysis. The development of city-scale climate monitoring systems, expected to be widespread by 2055, will enable precise tracking of urban heat islands, air quality, and local weather patterns. These systems will become increasingly important for urban planning and adaptation, helping cities develop responsive strategies to address climate challenges in real-time.

The marine component of real-time climate analysis is undergoing significant evolution. Advanced ocean monitoring systems, incorporating autonomous vehicles and floating sensors, provide continuous data on ocean temperatures, currents, and chemistry. By 2060, these systems are expected to achieve near-complete coverage of the world's oceans, offering unprecedented insights into marine climate dynamics and their influence on global weather patterns.

Data integration and processing capabilities are advancing rapidly, with new architectures being developed to handle the massive volume of climate data generated each second. The implementation of edge computing and distributed processing networks allows for more efficient data handling and analysis, reducing latency and improving the timeliness of climate insights. By 2050, these systems are expected to process and analyze climate data in microseconds, enabling truly real-time response capabilities.

The democratization of climate data access through real-time analysis platforms is transforming public engagement with climate science. Interactive visualization tools and user-friendly interfaces make complex climate data accessible to policymakers, researchers, and the general public. This accessibility is crucial for building public understanding and support for climate action and enabling more informed decision-making at all levels of society.

Real-time climate analysis systems are revolutionizing weather prediction capabilities. Integrating historical climate data with real-time observations enables more accurate forecasting across multiple timescales. By 2055, these systems are expected to achieve accuracy rates exceeding 95% for short-term weather predictions and significantly improve accuracy for seasonal and annual forecasts.

Agricultural applications of real-time climate analysis are becoming increasingly sophisticated. Advanced systems provide farmers with immediate insights into local weather patterns, soil conditions, and crop stress factors. By 2060, these systems are expected to incorporate predictive modeling capabilities that can anticipate crop yields and potential climate-related risks months in advance, enabling more resilient agricultural practices.

The role of AI in real-time climate analysis continues to expand. Machine learning algorithms are being developed that can automatically identify and classify weather patterns, track storm systems, and predict extreme weather events with increasing accuracy. These AI systems are expected to achieve human-level expertise in climate analysis by 2050, augmenting human capabilities in climate science and forecasting.

Global coordination of real-time climate analysis systems is improving through international partnerships and data-sharing agreements. The development of standardized protocols and interfaces enables seamless data integration from different sources and regions. By 2060, we anticipate the emergence of a truly global climate analysis network, providing comprehensive, real-time insights into Earth's climate system.

Edge Computing in Climate Monitoring

Edge computing is revolutionizing the landscape of climate monitoring by bringing computational power directly to the source of data collection. This paradigm shift in data processing and analysis represents a fundamental change in how we approach climate monitoring and prediction through 2050-2060. The integration of edge computing with climate monitoring systems addresses critical challenges in data latency, bandwidth limitations, and real-time analysis capabilities.

The current implementation of edge computing in climate monitoring focuses on distributed processing nodes positioned strategically across monitoring networks. These edge devices perform initial data processing, filtering, and analysis at the point of collection, significantly reducing the volume of raw data that needs to be transmitted to central processing facilities. By 2040, we expect to see the deployment of advanced edge processors capable of performing complex climate model calculations locally, enabling immediate response to changing environmental conditions.

A key development in edge computing for climate monitoring is the emergence of smart sensors equipped with integrated processing capabilities. These devices combine traditional sensing equipment with powerful microprocessors and artificial intelligence algorithms, enabling sophisticated data analysis at the source. The next generation of these smart sensors, anticipated by 2045, will incorporate neural processing units specifically designed for environmental data analysis, allowing for more sophisticated pattern recognition and anomaly detection at the edge.

The application of edge computing in remote climate monitoring stations is particularly significant. In areas with limited connectivity or harsh environmental conditions, edge computing systems provide crucial processing capabilities that ensure continuous data collection and analysis even when communication with central facilities is compromised. By 2050, these remote stations are expected to achieve near-complete autonomy, capable of making sophisticated decisions about data collection and analysis strategies based on local conditions.

Maritime climate monitoring is being transformed by edge computing capabilities integrated into buoys, autonomous surface vessels, and underwater monitoring systems. These edge-enabled platforms can process complex oceanographic data locally, providing immediate insights into marine climate conditions without the need for constant communication with shore-based facilities. The evolution of these systems through 2055 will enable more comprehensive monitoring of ocean-atmosphere interactions, crucial for understanding global climate patterns.

The integration of edge computing with satellite-based climate monitoring systems represents another frontier in climate science. On-board processing capabilities in weather and climate satellites are becoming increasingly sophisticated, allowing for preliminary data analysis and pattern recognition before transmission to ground stations. By 2060, we anticipate satellite systems capable of performing complex climate model calculations in orbit, providing real-time insights into global atmospheric patterns.

Energy efficiency in edge computing systems is advancing rapidly, with new architectures being developed specifically for environmental monitoring applications. Low-power processing units combined with renewable energy sources enable continuous operation in remote locations. The next generation of edge computing systems, expected by 2050, will achieve near-zero environmental impact while providing significantly increased processing capabilities.

The role of AI at the edge is becoming increasingly important for climate monitoring. Edge-based AI systems can perform sophisticated analysis of local climate data, identifying patterns and trends that traditional monitoring approaches might miss. These systems are expected to achieve human-level expertise in local climate analysis by 2055, providing crucial support for climate scientists and environmental managers.

Data security and reliability are enhanced through edge computing implementations in climate monitoring networks. Distributed processing reduces the risk of data loss or corruption during transmission, while edge-based security protocols protect sensitive climate data from unauthorized access. Future developments in edge security, anticipated by 2060, will include quantum-resistant encryption implemented directly at the sensor level.

Urban climate monitoring particularly benefits from edge computing capabilities. Smart city infrastructure equipped with edge processing can provide real-time analysis of urban heat islands, air quality, and local weather patterns. The evolution of these systems through 2050 will enable more effective urban climate adaptation strategies, with immediate response capabilities to changing environmental conditions.

The standardization of edge computing protocols in climate monitoring is crucial for system interoperability and data integration. International efforts are underway to develop common standards for edge-based climate monitoring systems, ensuring compatibility across different platforms and regions. By 2060, we expect to see fully standardized edge computing architectures specifically designed for climate monitoring applications.

Research and development in edge computing for climate monitoring continue to advance rapidly. New processing architectures, more efficient algorithms, and improved integration with existing climate monitoring systems are being developed. The next decade will see significant improvements in edge computing capabilities, enabling more sophisticated climate analysis and prediction at local and regional scales.

Additional Quantum Computing Notes

In previously detailed environmental areas, quantum computing significantly assists scientists by providing unprecedented computational power for modeling and analyzing intricate climate systems. As we anticipate the period between 2050 and 2060, the integration of quantum computing with climate science holds the potential to revolutionize our comprehension of global climate dynamics and augment our capacity to forecast future climate scenarios with unparalleled accuracy.

Quantum machine learning applications in climate science are showing particular promise. These hybrid systems combine quantum computing capabilities with advanced AI algorithms to identify patterns in climate data that would be impossible to detect using traditional methods. The evolution of these systems through 2050 will enable more sophisticated analysis of climate feedback loops and tipping points, which is crucial for understanding potential future climate scenarios.

The application of quantum computing to atmospheric chemistry calculations represents another crucial advancement. Quantum systems can efficiently simulate molecular interactions in the atmosphere, providing more accurate models of greenhouse gas behavior and chemical weather predictions. By 2060, these capabilities are expected to enable precise modeling of atmospheric composition changes and their impacts on global climate.

Ocean modeling benefits significantly from quantum computing capabilities. The complex fluid dynamics of ocean currents and their interaction with atmospheric systems can be more accurately simulated using quantum algorithms. Future quantum systems will enable more precise modeling of ocean heat content and circulation patterns, crucial factors in global climate regulation.

Cloud formation and precipitation patterns are particularly challenging aspects of climate modeling that quantum computing is helping to address. The ability to simultaneously process multiple atmospheric variables allows for a more accurate simulation of cloud dynamics and rainfall patterns. By 2055, quantum-enabled weather prediction systems are expected to achieve unprecedented accuracy in forecasting precipitation events across multiple timescales.

The economic implications of quantum computing in climate science are significant. More accurate climate predictions enable better planning for climate adaptation and mitigation strategies, potentially saving trillions of dollars in climate-related damages. The investment in quantum computing infrastructure for climate science is expected to increase dramatically through 2050 as the technology demonstrates its value in improving climate predictions.

Data integration between quantum and classical computing systems is becoming increasingly sophisticated. Hybrid quantum-classical architectures are being developed to optimize the strengths of both systems, enabling more efficient processing of climate data. By 2060, we anticipate seamless integration between quantum and classical systems in climate modeling applications.

Research into new quantum computing architectures specifically designed for climate applications continues to advance. These specialized systems will be optimized for the unique requirements of climate modeling, including long coherence times and high-fidelity quantum operations. The development of these systems through 2050 will enable more sophisticated climate simulations and predictions.

Global Climate Models by Federated Learning.

The integration of federated learning into global climate modeling represents a transformative approach to environmental science. This fusion of distributed computing capabilities with the urgent need for precise climate predictions addresses several critical challenges in climate modeling. Moreover, it presents novel opportunities for collaboration and data utilization across geographical and institutional boundaries.

Traditionally, global climate models (GCMs) relied on centralized computing infrastructure, where substantial amounts of data from diverse sources were aggregated and processed in powerful supercomputers. While this approach has yielded substantial insights into climate patterns and future projections, it has encountered several limitations. These include data privacy concerns, computational bottlenecks, and the challenge of effectively incorporating local variations and expertise.

The proliferation of diverse data types originating from disparate sources necessitates the development of sophisticated models to effectively process this data. The intricacies of climate systems necessitate models that incorporate satellite observations, weather stations, ocean buoys, and local environmental sensors. These data sources are often managed by distinct organizations across various jurisdictions, each adhering to its own data privacy regulations and frameworks. Consequently, traditional centralized approaches encounter challenges in accommodating these constraints while maintaining the comprehensive scope required for accurate climate modeling.

A paradigm shift was required to address the challenge of collaborative model training without centralizing raw data. This approach, known as "federated learning," enables multiple parties to contribute to a shared global climate model while maintaining the privacy of their sensitive data. The process typically involves several key steps. Each participating institution trains a model component using local climate data and computational resources.

This could include regional weather patterns, oceanographic data, or atmospheric measurements specific to their geographic area.

The locally trained models' parameters are shared and aggregated to update a global model rather than sharing the raw data itself. This aggregation process preserves privacy while incorporating insights from all participating entities. The improved global model is then distributed back to all participants, who can further refine it using their local data, creating an iterative improvement cycle.

Federated learning facilitates the integration of local expertise and specialized knowledge pertaining to regional climate patterns, thereby enhancing data privacy, which may be compromised in a centralized approach. Additionally, it alleviates the computational burden on any single facility by distributing the processing across multiple nodes.

Implementing federated learning in climate modeling presents unique technical challenges that require innovative solutions. One primary consideration is the heterogeneity of data sources and computational resources across participating institutions. Climate data comes in various formats, resolutions, and measurement frequencies, necessitating sophisticated harmonization techniques.

The architecture of federated learning systems for climate modeling commences with the design of the model architecture. The climate model must be meticulously crafted to facilitate federated training, carefully considering which components can be trained locally versus globally. This may entail hierarchical structures that amalgamate local weather patterns with global circulation models.

To facilitate the efficient sharing of model parameters, protocols are necessary to effectively manage bandwidth limitations, given the substantial parameter spaces associated with climate models. The model will interact with diverse sources of varying quality and employ intricate aggregation methods. These algorithms must demonstrate robustness to statistical heterogeneity and potential biases inherent in local datasets.

The adoption of federated learning in climate modeling entails far-reaching implications. It must adhere to local privacy regulations, which may necessitate the anonymization of certain parameters without compromising the analysis's quality. Data quality verification and protection against adversarial attacks or obfuscation during aggregation are essential. Secure communication channels software or, in certain cases, hardware-controlled solutions may be required. Access control for both ingress and egress must be a robust component of the infrastructure.

The integration of diverse data sources and local expertise can enhance model accuracy and diminish prediction uncertainties, particularly for regional climate patterns. Distributed computing resources can be utilized more effectively, enabling smaller institutions to make substantial contributions to global climate research.

More precise and locally pertinent climate predictions can guide improved adaptation and mitigation strategies at both regional and global scales.

Emerging technologies such as edge computing and 5G networks can further enhance the efficiency of distributed model training. Integration with other AI techniques, such as reinforcement learning for parameter optimization or computer vision for satellite data analysis, could expand the capabilities of these systems. More efficient aggregation algorithms specifically designed for climate data. Creating standardized protocols for model sharing and validation and integrating real-time data streams for continuous model updating would significantly enhance the analysis and projection.

The ongoing advancement and refinement of federated learning methodologies in climate modeling will undoubtedly play a pivotal role in our comprehension and response to climate change challenges. By facilitating more inclusive and comprehensive climate modeling endeavors, this technology contributes to the global scientific community's capacity to address one of the most pressing challenges of our era.

Impact on Various Areas

Wood Processing and Manufacturing

Despite utilizing renewable resources, the woodworking industry contributes substantially to global greenhouse gas emissions throughout its various production and processing stages. This analysis evaluates the climate impact of wood processing, encompassing the extraction of raw materials and the production of finished products, focusing on the environmental consequences associated with each stage of the value chain.

Carbon Storage and Deforestation

The initial phase of wood production entails intricate disruptions to established forest ecosystems that extend far beyond the immediate impact of tree removal. When forests are cleared for timber, their capacity as carbon sinks is drastically reduced, initiating a cascade of environmental effects that can persist for decades.

Mature forests represent sophisticated carbon storage systems operating across multiple integrated layers. The canopy layer, composed of mature trees, stores substantial amounts of carbon in trunks, branches, and leaves. Below this, the mid-story layer of younger trees and larger shrubs provides continuous carbon sequestration, while the understory layer of small plants and saplings contributes to ongoing carbon capture. The forest floor consists of decomposing organic matter that gradually builds carbon-rich soil, supported by extensive underground root systems that store carbon and maintain soil structure. Throughout all these layers, intricate communities of soil organisms regulate carbon cycling in a delicate equilibrium.

The disruption of these systems through logging creates multiple pathways for carbon release. Immediate carbon emissions occur through various mechanisms: trees being cut release their stored carbon directly, while disturbed vegetation and forest floor materials undergo accelerated decomposition. The exposure of soil organic matter leads to increased oxidation, and damaged root systems release their stored carbon. Perhaps most significantly, the disruption of mycorrhizal networks that store and transport carbon throughout the forest system creates long-lasting impacts on carbon sequestration capabilities.

The long-term reduction in carbon storage capacity manifests in various ways. The decreased canopy density significantly reduces the forest's photosynthetic capacity, while soil compaction from heavy equipment limits future tree growth potential. The loss of seed sources affects natural regeneration patterns, altering microclimates and changing decomposition rates throughout the system. These changes in hydrology have lasting impacts on soil carbon retention.

The disruption extends to essential ecosystem services as well. The reduced habitat connectivity impedes pollinator movements and species distribution, while altered light conditions alter the composition of understory vegetation. Modified soil chemistry influences nutrient cycling throughout the system, and disturbed watershed functions affect overall soil stability. These altered species interactions have significant implications for forest resilience and recovery potential.

Turning raw logs into usable lumber is a big energy hog that leaves much waste behind. Modern sawmills are more efficient than old-school ones but still make a big mess for the environment. They use a ton of electricity to power the sawing machines and heat up the kilns for drying. And the waste management process has become another way to pollute the air. Plus, transporting logs between different stages adds to the carbon footprint. The kiln drying process is especially energy-intensive, often using fossil fuels to keep the temperature just right for a long time.

The conversion of raw logs into usable lumber entails substantial energy consumption and generates substantial waste. Modern sawmills, while more efficient than historical operations, still contribute significantly to climate change. The process demands extensive electricity consumption for sawing equipment, while thermal energy is required for kiln drying operations. Waste management processes create additional environmental burdens, and the transportation between various processing stages adds to the overall carbon footprint. The kiln drying process is particularly energy-intensive, commonly relying on fossil fuels to maintain consistent temperatures over extended periods necessary for proper wood conditioning.

Chemical treatments used to protect wood from decay and insects have some serious environmental drawbacks. The process of making chemical preservatives releases a lot of pollution, and the treatments themselves use a lot of energy. Chemical reactions during treatment also release more pollution, and getting rid of the contaminated waste is a never-ending problem. These treatments do make wood last longer, but they often use toxic chemicals that cause problems beyond just the climate.

The production of engineered wood products and panels generates substantial emissions through various processing stages. Medium Density Fiberboard (MDF) production entails energy-intensive fiber processing, followed by adhesive production and application. The hot pressing operations consume substantial energy, and edge finishing and surface treatment processes further contribute to the environmental impact.

Similarly, the production of plywood and particleboard follows similar energy-intensive patterns. The process commences with veneer drying and processing, continues through adhesive manufacturing and application, and necessitates substantial energy for pressing and heating cycles. The finishing operations conclude the emission-generating sequence.

Furniture production is a process that involves many steps that release emissions. From cutting and shaping the wood to applying paint and varnish, there are many ways that furniture production can hurt the environment.

The Packaging and Movement of Wood

Product protection during transportation necessitates the use of additional materials that contribute to environmental degradation. Traditional packaging methods heavily rely on plastic wrapping, cardboard packaging, protective foam, and strapping and pallets. These materials frequently become waste after a single use, contributing to both direct emissions and broader environmental challenges.

The transportation of wood products through the supply chain generates substantial emissions through a complex network of transportation and handling operations. Primary transportation from forest to processing facilities involves heavy logging trucks operating on unpaved forest roads, with fuel consumption increased by rough terrain and heavy loads. Additional equipment for loading and unloading, maintenance vehicles, and support transportation all contribute to emissions. The construction and maintenance of temporary roads and bridges for access add another layer of environmental impact.

Inter-facility movement generates additional environmental burdens through extended transportation between processing facilities and multiple handling operations at transfer points. Climate-controlled storage requirements, loading and unloading equipment, and fuel consumption for yard operations contribute to the carbon footprint. Emissions from staging and sorting operations further exacerbate the environmental impact.

Wood products continue to impact the environment throughout their lifecycle. Finishes on wood surfaces release Volatile Organic Compounds (VOCs) into indoor spaces, while adhesives used in construction release chemicals over time. Cleaning wood surfaces creates particulate matter, and refinishing processes release more chemicals into the air. These ongoing emissions lead to indoor air quality issues and broader environmental problems.

Manufacturing distribution encompasses regional distribution center operations, cross-docking facilities, and extensive material handling infrastructure. Climate control requirements for specific products, vehicle emissions from distribution fleets, and energy consumption for logistics operations contribute to the environmental burden.

Retail and consumer delivery networks create additional impacts through last-mile delivery operations, urban distribution challenges, and multiple handling points within the retail chain. Customer return transportation, packaging waste management, and storage facility operations further increase the carbon footprint.

Global supply chain impacts extend to international shipping emissions, port facility operations, and customs and inspection delays. Intermodal transfer emissions, additional packaging requirements, and extended storage needs create additional environmental challenges.

Temperature-controlled transport introduces additional complexity, necessitating additional fuel consumption for climate control and specialized equipment. Enhanced packaging requirements, increased maintenance operations, backup systems for critical loads, and emergency response capabilities all contribute to the overall environmental impact.

Protecting products during transit often involves using materials that can harm the environment. Traditional packaging methods heavily rely on plastic wrapping, cardboard boxes, protective foam, and strapping and pallets. These materials usually end up as waste after just one use, causing both direct emissions and bigger environmental problems.

The maintenance of wood products during their lifespan generates a continuous stream of environmental impacts. Regular cleaning necessitates the use of chemical products and polishes that emit both production and disposal-related emissions. Refinishing materials further increases the chemical burden, while repair products and replacement components necessitate new manufacturing processes and associated emissions. The cumulative effect of these maintenance requirements significantly extends the product's environmental footprint beyond its initial production impact.

Wood products don't just end when they're used up. They create a lot of problems for the environment. When we throw them away in landfills, they break down and release methane, a gas that traps heat in the atmosphere. And the chemicals we use to treat them can leak into the soil and water, polluting them. Plus, the transportation of all that waste adds to the pollution. And the land we need to store it all takes up a lot of space, which can hurt the environment.

When we burn wood, we get some energy back, but it comes at a cost. We release carbon dioxide and other pollutants into the air, and the ash we get from burning can be toxic. And even though we try to make sure the process is as efficient as possible, it's not always enough to make up for all the environmental damage we're causing.

Wood recycling can be a good thing for the environment, but it's not without its drawbacks. The process uses a lot of energy, and the sorting and cleaning can be resource-intensive. Unfortunately, non-recyclable parts end up in landfills, which can negate some of the benefits of recycling.

Cumulative Impact

Wood products have a big impact on the climate. They release carbon dioxide during processing and transportation, and they break down over time, releasing more carbon.

The environmental impact of wood processing extends far beyond immediate climate effects. Forest ecosystem stability faces ongoing challenges from resource extraction, while biodiversity preservation necessitates careful management of harvesting practices. Soil quality and erosion control have become increasingly important considerations in managed forests, and water system health requires constant monitoring and protection measures.

What can the industry do?

The wood processing industry continues developing methods to reduce climate impact through various technological and procedural improvements. Sustainable forestry practices have evolved to include more selective harvesting techniques that minimize ecosystem disruption. Reforestation programs work to maintain forest coverage, while forest certification systems help ensure responsible management practices. Biodiversity preservation measures have become increasingly integrated into forest management plans.

Process improvements within manufacturing operations focus on implementing energy-efficient equipment and waste reduction systems. Adopting clean energy sources helps reduce direct emissions, while closed-loop water systems minimize resource consumption and environmental impact. These innovations represent ongoing efforts to balance production requirements with environmental responsibility.

Consumer choices can make a big difference in how the industry works and how the environment is affected. When products that are certified as sustainable are picked, this helps to make sure that forests are managed responsibly. When products made locally are selected, this helps reduce the amount of pollution that comes from transportation. Taking care of wood products so they last longer means they have not to be replaced as often, and supporting recycling programs helps keep the environment clean at the end of their life.

The woodworking industry's effects on the environment go beyond just cutting down trees. Every step in making wood products adds up to a lot of environmental damage, from how we manage forests to how we get rid of old wood. Knowing these effects is important for finding ways to reduce them and making smart choices as consumers.

Wood is a renewable resource that could be better for the environment than other materials, but the way we process it and the industry practices we have right now are really bad for the climate. We need everyone involved in the woodworking industry to work together to make things better, from how we take care of forests to how we make and dispose of wood products.

We can make progress in reducing the environmental impact of wood processing by continuing to invent new technologies, getting the government to support us, and educating people about the issue. As the industry changes, it's important to find a balance between making things quickly and being responsible for the environment.

Steel Manufacturing

The environmental impact of steel manufacturing extends far beyond the visible emissions from plant stacks. Each stage of production, from raw material extraction through end-of-life processing, contributes to climate change through direct and indirect emissions. Although technological advancements continue to reduce the environmental impact per ton of steel produced, the industry's substantial scale ensures that it remains one of the most significant contributors to global greenhouse gas emissions.

Steel production is a major contributor to global climate change, causing about 7-9% of direct emissions from burning fossil fuels. This study looks at the whole process of making steel, from getting the raw materials to throwing away the finished product, to show how all these steps affect the environment and cause climate change.

The environmental impact of steel production commences well before any metal is forged. Iron ore mining, a prerequisite for steel production, causes extensive environmental disruption across vast geographical regions. Open-pit mining operations remove surface vegetation and topsoil, releasing stored carbon from plant matter and soil organic compounds. The exposure of deeper soil layers to air accelerates the oxidation of organic materials, resulting in the release of additional carbon dioxide into the atmosphere.

The extraction process necessitates continuous operation of heavy machinery, which consumes substantial amounts of diesel fuel and contributes significantly to emissions. Drilling and blasting operations, necessary for accessing deeper ore deposits, release particulate matter and gases into the atmosphere. The removal of overburden, comprising rock and soil above the ore body, further disrupts the environment and necessitates energy-intensive handling and disposal processes.

Coal mining, crucial for producing coking coal used in blast furnaces, presents its own severe environmental challenges. Underground mining operations release methane, a greenhouse gas significantly more potent than carbon dioxide. Ventilation systems in underground mines continuously expel this methane into the atmosphere. Surface mining for coal disrupts natural carbon sinks and often leads to the destruction of forests and other vegetation, further diminishing the Earth's capacity to absorb atmospheric carbon dioxide.

The transportation of raw materials generates additional environmental impacts. Massive ore carriers and bulk transport vessels powered by heavy fuel oil transport iron ore and coal across global shipping routes. These vessels, frequently operating on the lowest grade of fuel oil, release substantial amounts of sulfur dioxide and carbon dioxide. The loading and unloading operations at ports necessitate additional energy-intensive equipment and infrastructure.

The initial processing of iron ore involves crushing and grinding operations that consume substantial electrical energy. The beneficiation process, which concentrates the iron content through magnetic separation or flotation, requires substantial water resources and chemical additives. The disposal of tailings—the waste material from this process—creates large impoundments that can release greenhouse gases through the decomposition of organic matter and chemical reactions.

The transformation of coal into coke is a pivotal stage in steel production, characterized by the extreme heating of coal in the absence of oxygen. This process eliminates volatile compounds, resulting in pure carbon fuel essential for blast furnace operations. However, coking also generates substantial quantities of greenhouse gases and other pollutants directly into the atmosphere.

Coking plants operate continuously, consuming substantial energy to maintain the elevated temperatures required for the process. The cooling of hot coke generates thermal emissions and necessitates substantial water resources. The quenching process, which involves cooling hot coke with water or nitrogen, releases steam containing pollutants into the atmosphere. Additionally, handling and storing raw coal and finished coke generate dust emissions that contribute to local air quality degradation and climate impacts.

The blast furnace process is the main way steel is made, and it's the biggest source of pollution from the steel industry. Iron ore is heated up with coke as fuel and a reducing agent, releasing a lot of carbon dioxide. Even though modern blast furnaces are more efficient than old ones, they still need temperatures over 1,500 degrees Celsius, which uses a lot of energy.

Blast furnaces can't be easily turned on or off, so they need to keep running even when they're not making as much steel. The hot blast stoves that warm up the air before it goes into the furnace use more fossil fuels. The blast furnace gas, which is used to make electricity, has a lot of carbon monoxide and carbon dioxide that eventually go into the air.

The conversion of iron into steel through Basic Oxygen Furnace (BOF) or Electric Arc Furnace (EAF) processes generates additional environmental impacts. The BOF process, which utilizes high-purity oxygen to reduce the carbon content of iron, directly releases carbon dioxide through the chemical reaction. The high temperatures required for this process necessitate substantial energy input, while the handling of hot metal and slag results in thermal emissions.

Electric arc furnaces, while generally perceived as more environmentally friendly due to their capacity to process recycled steel, still contribute significantly to climate change through their substantial electricity consumption. In regions where electricity generation predominantly relies on fossil fuels, the indirect emissions from EAF operations can be substantial. The high-temperature electric arcs also lead to electrode consumption, necessitating continuous replacement and generating additional manufacturing emissions.

The transformation of raw steel into finished products entails multiple heating and cooling cycles, each contributing to the overall climate impact. Hot rolling operations necessitate reheating furnaces that consume fossil fuels to elevate steel temperatures exceeding 1,200°C. Subsequent cooling processes release thermal energy into the environment, while scale formation and removal generate additional waste and necessitate energy-intensive handling.

Cold rolling processes, while not requiring the extreme temperatures of hot rolling, consume substantial electrical energy to achieve the required deformation of the steel. The multiple passes required for attaining final dimensions further amplify the energy consumption. Surface treatment processes, such as pickling and galvanizing, involve chemical reactions and heating operations that generate both direct emissions and hazardous waste, requiring specialized handling.

The management of waste materials and by-products generated during steel production poses ongoing environmental challenges. Slag handling and processing necessitate energy-intensive equipment and result in dust emissions. While some slag finds applications in construction, its transportation and processing further contribute to emissions. The treatment of wastewater from diverse processes consumes energy and chemicals, while the disposal of contaminated sludge can release greenhouse gases through decomposition.

The handling of captured dust and fumes from various processes necessitates sophisticated filtration systems that consume electrical energy. Disposing or recycling these materials entails additional transportation and processing steps, each contributing to the overall carbon footprint of steel production. The storage and handling of hazardous wastes from surface treatment processes pose risks of fugitive emissions and necessitate stringent environmental controls.

The steel production process has a big environmental impact. Cooling water systems, which are crucial for many processes, need pumping stations and treatment facilities that use a lot of electricity. Compressed air systems, which are used all over steel plants for controlling things and moving materials, are one of the biggest electricity users in many places.

Material handling systems, like conveyor networks, cranes, and mobile equipment, keep moving raw materials and products around the facility all the time. Taking care of this equipment uses more energy and makes waste that needs to be disposed of. The lighting and ventilation of big industrial buildings also use a lot of energy, and the maintenance of roads and other infrastructure creates more emissions because of construction and the use of materials.

Steel products travel the world, causing a lot of pollution. Big ships carry steel across oceans, while trucks and trains move it around locally. The way we handle steel at ports and distribution centers also adds to the pollution. And the packaging we use to protect steel during shipping creates extra waste and pollution.

Although steel remains theoretically recyclable, the practical process of recovering and processing end-of-life steel products generates substantial emissions. The collection and sorting of scrap necessitate energy-intensive equipment and transportation. The shredding and processing of automotive and industrial scrap consume substantial electrical energy, resulting in noise and dust emissions. The separation of steel from other materials in complex products often necessitates additional processing steps that further contribute to the overall environmental impact.

Plastics and Petrochemical Industry

The environmental impact of petrochemical and plastic manufacturing encompasses the entire value chain, from raw material extraction to end-of-life disposal. The industry's reliance on fossil fuels as both feedstock and energy source poses fundamental challenges in mitigating climate change. Despite technological advancements and efficiency gains that have reduced emissions per unit of production, the escalating global demand for plastic products has resulted in a persistent increase in the overall environmental impact.

The environmental impact starts with the extraction of fossil fuels, like oil and natural gas. These fuels are used to make petrochemicals and provide energy. Offshore drilling needs huge platforms that use a lot of energy to run, while onshore extraction often uses hydraulic fracturing to release methane, which escapes into the air. The drilling process itself uses energy-intensive equipment that runs all the time, making a lot of greenhouse gas emissions.

The extraction process often involves flaring extra natural gas, which releases carbon dioxide and methane directly into the atmosphere. This is especially common in areas without enough gas collection infrastructure, and it contributes a lot to the industry's climate impact. The transportation of crude oil through pipelines and ships creates more emissions because of the energy used and occasional leaks.

Crude oil refining constitutes the initial significant processing step in the petrochemical production process. The distillation process entails heating crude oil to exceptionally high temperatures, necessitating substantial energy typically sourced from fossil fuels. The fractional distillation towers operate continuously, maintaining temperature gradients that separate distinct hydrocarbon components. This continuous operation generates both direct emissions from fuel combustion and indirect emissions stemming from electricity consumption.

The refining process generates substantial volumes of greenhouse gases through various chemical reactions and energy-intensive separation processes. Catalytic cracking units, indispensable for breaking down heavy hydrocarbons into lighter components, operate at elevated temperatures and pressures, necessitating substantial energy input. The regeneration of catalysts involves burning off carbon deposits, thereby releasing additional carbon dioxide into the atmosphere.

The conversion of refined petroleum products into fundamental petrochemicals entails intricate chemical processes that generate substantial emissions. Steam cracking, the primary process for producing ethylene and propylene, necessitates temperatures exceeding 800°C. This extreme heat requirement typically relies on fossil fuel combustion, resulting in significant carbon dioxide emissions. The subsequent rapid cooling required after cracking consumes additional energy and often necessitates the utilization of refrigeration systems that may release hydrofluorocarbons.

The production of aromatic compounds through catalytic reforming further exacerbates environmental impacts. Operating at elevated temperatures and pressures, this process generates direct emissions from chemical reactions and indirect emissions from energy consumption. The separation and purification of various aromatic compounds necessitate multiple distillation steps, each consuming substantial energy and potentially releasing volatile organic compounds.

The process of turning basic petrochemicals into plastics is another big contributor to emissions. The polymerization process needs super-precise temperature control and often involves high-pressure conditions, which means it uses a lot of energy. The different types of plastics, like polyethylene and polyvinyl chloride, each have their own special catalysts and conditions that affect the environment.

The processes of extruding and forming raw polymers into useful products use a lot of electricity. The heating and cooling cycles for these processes create thermal emissions, and handling additives and stabilizers adds chemical risks. Making specialty plastics often requires more complicated processes and extra chemical treatments, which makes it even more environmentally harmful.

The incorporation of additives into plastic materials presents additional environmental concerns. The production of colorants, stabilizers, plasticizers, and flame retardants necessitates distinct chemical manufacturing processes, each contributing its own emissions. The compounding process, which combines these additives with base polymers, necessitates mechanical energy for mixing and temperature control for optimal incorporation.

The handling and storage of various chemical additives pose risks of fugitive emissions and necessitate sophisticated containment systems. Many additives are persistent organic pollutants that contribute to long-term environmental degradation. The production of master batches and specialty compounds entails additional processing steps that consume energy and potentially release volatile compounds.

Plastic bag production

The transformation of plastic resins into finished products entails a series of manufacturing processes that contribute to climate change. Injection molding machines, which are crucial for this process, consume substantial electrical energy to maintain optimal temperatures and pressures. The rapid heating and cooling cycles necessary for efficient production generate thermal emissions, while the operation of hydraulic systems and auxiliary equipment further elevates the energy demand.

Blow molding operations, which are essential for producing hollow items such as bottles, necessitate compressed air systems that consume considerable electricity. The production of films and sheets through extrusion processes involves continuous heating and cooling cycles. The trimming and finishing of plastic products generate waste that necessitates collection and either disposal or recycling, thereby adding to the environmental impact.

The transportation of materials throughout the production process generates substantial transportation-related emissions. Raw materials and finished products frequently travel extensive distances between processing facilities, typically relying on fossil fuel-powered transportation. The global nature of the petrochemical industry necessitates the frequent movement of products between continents, resulting in significant shipping emissions.

The storage and handling of materials at various stages require specialized facilities equipped with temperature and humidity control systems that consume energy continuously. The packaging and protection of plastic products for shipment often involve additional plastic materials, perpetuating a cycle of environmental impact.

The management of production waste poses a substantial environmental challenge. The handling of off-specification materials and production scrap necessitates energy consumption for collection, sorting, and reprocessing. While certain materials can be recycled internally, many require disposal through incineration or landfilling, both of which generate additional greenhouse gas emissions.

The release of microplastics during production processes results in persistent environmental contamination. These particles infiltrate water systems through industrial wastewater and air through ventilation systems. The degradation of plastic materials in the environment releases greenhouse gases and other pollutants, contributing to long-term climate impacts.

The environmental impact of plastic products extends throughout their lifecycle, including their use and eventual disposal. Many plastic products emit Volatile Organic Compounds (VOCs) throughout their existence, contributing to atmospheric pollution. The degradation of plastics in the environment releases methane and ethylene, both potent greenhouse gases. The accumulation of plastic waste in marine environments exacerbates environmental stresses that may disrupt global carbon cycles.

Plastic Waste

The disposal of plastic products presents substantial challenges. Landfilling results in slow decomposition that releases greenhouse gases over extended periods. Incineration, while potentially generating energy, releases carbon dioxide and potentially hazardous compounds. Recycling, while preferable, necessitates energy for collection, sorting, cleaning, and reprocessing, often resulting in downcycled products of inferior quality.

Bio-based and biodegradable plastics have both advantages and disadvantages when it comes to the environment. They can help us reduce our reliance on fossil fuels, but they're often made using farming practices that release their own emissions. And the process of making these materials can use more energy than making traditional plastics from oil, which might cancel out some of the environmental benefits.

Carbon capture and storage technologies could help us reduce the emissions from the factories that make these plastics, but it would cost a lot of money and energy to set up. And chemical recycling technologies, which could help us deal with hard-to-recycle plastics, are also energy-intensive and could release more emissions.

> *Addressing these challenges necessitates a multifaceted approach that integrates technological advancements, regulatory frameworks, and alterations in consumer behavior.*

The development of renewable feedstocks, enhanced recycling technologies, and more efficient manufacturing processes present viable avenues for reduction. Nevertheless, the fundamental transformation required to substantially mitigate the industry's climate impact remains a pivotal challenge for future generations.

Cement Production

Cement production is a major contributor to global climate change, accounting for about 8% of all CO_2 emissions. This study looks at the whole environmental impact of cement making, from getting the raw materials to using the final product, to show how this important industry affects our planet's climate.

The environmental impact of cement production commences with limestone quarrying, which fundamentally alters landscapes and disrupts natural carbon sinks. The extraction process necessitates continuous operation of heavy machinery to remove vast quantities of limestone and clay. These operations typically involve drilling, blasting, and crushing activities that consume substantial amounts of diesel fuel while releasing particulate matter into the atmosphere.

Cement Production

The removal of surface vegetation and topsoil during quarrying eliminates natural carbon sequestration capacity. The exposure of deeper geological layers accelerates the weathering of minerals, potentially releasing additional carbon dioxide. The construction of quarry faces and access roads necessitates extensive earthmoving operations that generate significant emissions from fuel consumption and soil disturbance.

The initial processing of raw materials entails a series of crushing and grinding stages that consume substantial amounts of electrical energy. Primary and secondary crushers reduce limestone blocks to suitable sizes, while ball mills and vertical roller mills grind the material into a fine powder. This grinding process, crucial for the proper formation of clinker, is one of the most energy-intensive stages in cement production.

The preparation of raw meals necessitates precise blending of limestone, clay, and corrective materials such as iron ore and bauxite. The handling and storage of these materials generate dust emissions that contribute to local air quality degradation. The homogenization process requires additional energy for mixing and pneumatic transfer systems that operate continuously to ensure optimal material flow.

The pyro-processing stage, the core of cement production, involves transforming raw materials into clinker through extreme temperature exposure. This process commences in the preheater tower, where cyclones utilize hot exhaust gases to heat the raw meal. The counter-current flow of materials and gases enhances energy efficiency, albeit necessitating substantial heat input from fossil fuel combustion.

The rotary kiln stands as the primary source of direct CO_2 emissions in cement production. The chemical decomposition of limestone ($CaCO_3$) into lime (CaO) and carbon dioxide occurs at temperatures of approximately 900°C, releasing CO_2 as an inherent byproduct of the process. The kiln must maintain temperatures exceeding 1,450°C to form clinker minerals, necessitating continuous combustion of fossil fuels that generates additional CO_2 emissions.

The extreme temperatures required for clinker formation typically depend on fossil fuels such as coal, petroleum coke, and other carbon-intensive sources. While alternative fuels like waste materials and biomass are increasingly being utilized to supplement traditional fuels, the industry remains heavily dependent on these energy sources. The preparation and handling of fuels further contribute to environmental impacts through grinding, storage, and transportation operations.

The combustion process itself generates various pollutants beyond carbon dioxide, including nitrogen oxides, sulfur compounds, and trace elements that contribute to broader environmental degradation. The high temperature requirement necessitates continuous kiln operation, as thermal cycling would result in excessive energy waste and potential equipment damage.

The rapid cooling of clinker from its peak temperatures is another energy-intensive process. Clinker coolers utilize substantial volumes of air to reduce the material temperature while simultaneously recovering some heat for utilization in other processes. The handling of hot clinker generates thermal emissions and necessitates robust equipment that consumes additional energy for operation.

The storage and transportation of clinker result in dust emissions that necessitate the collection and handling of these emissions through specialized systems. The maintenance of storage facilities and transfer equipment further contributes to the environmental impact by consuming energy and causing material wear. The processing of collected dust often necessitates the return of material to earlier stages of production, perpetuating a cycle of energy consumption.

The transformation of clinker into finished cement entails a subsequent series of intensive grinding operations. Ball mills or vertical roller mills combine clinker with gypsum and other additives, consuming substantial electrical energy to attain the desired particle fineness. The precise control of cement fineness necessitates sophisticated classification systems, which further contribute to the overall energy demand.

The handling and storing of grinding aids and cement additives present the potential for chemical emissions and necessitate stringent environmental controls. The packaging and loading of finished cement generate dust emissions that necessitate collection and handling. The maintenance of grinding equipment generates additional waste streams requiring proper disposal.

The extensive infrastructure necessary for cement production generates its own environmental footprint. The construction and maintenance of kilns, mills, storage silos, and handling systems require substantial material and energy inputs. The electrical systems required for plant operation often rely on fossil fuel-generated power, contributing to indirect emissions.

The ongoing development and maintenance of quarry infrastructure, including access roads, drainage systems, and processing facilities, necessitates continuous construction activity that generates additional emissions. The eventual decommissioning of cement plants and quarries presents further environmental challenges through demolition and site rehabilitation requirements.

The transportation of materials throughout the cement production process generates substantial transportation-related emissions. Raw materials typically travel considerable distances from quarries to production facilities, while finished cement requires distribution to concrete plants and construction sites. The substantial weight of cement products necessitates substantial fuel consumption, whether via truck, rail, or marine vessel.

Cement Transport

Although concrete production is a distinct industrial process, its utilization continues to generate environmental impacts beyond its manufacturing phase. The mixing and placement of concrete necessitate additional energy inputs, while the curing process can release further CO_2 through carbonation reactions. Furthermore, the demolition and disposal of concrete structures at the end of their lifespan generate additional emissions through mechanical processing and transportation.

Managing various waste streams generated during cement production poses ongoing environmental challenges. Kiln dust, collected from air pollution control systems, necessitates careful handling due to its concentrated levels of heavy metals and other pollutants. Disposing of refractory materials from kiln maintenance generates additional waste requiring specialized handling.

The treatment of process water generates sludge containing various contaminants that necessitate proper disposal. Additionally, the management of used equipment and maintenance materials generates additional waste streams that contribute to the overall environmental impact of cement production.

Cement production has a significant environmental impact, extending beyond the visible emissions from factories. Each step in the cement-making process contributes to climate change, both directly and indirectly. While progress is being made in reducing emissions, the industry's scale and complexity pose challenges to achieving substantial reductions.

Summary

T he continuation of our current trajectory without substantial alterations in our approach to nature conservation and climate change mitigation would result in profound and far-reaching transformations in our world. The cascading effects would permeate every facet of life on Earth, leading to the creation of a fundamentally different planet for future generations.

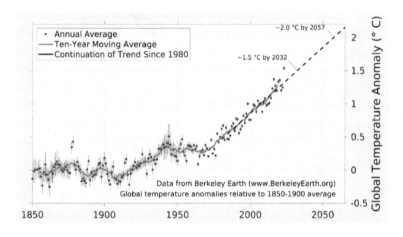

By the mid-century, coastal cities globally would face unprecedented challenges due to the accelerated rise in sea levels. Major metropolitan areas such as Miami, Shanghai, and Amsterdam would necessitate substantial infrastructure investments in sea walls and water management systems. Consequently, some low-lying neighborhoods would become uninhabitable. Island nations in the Pacific would witness the gradual disappearance of their territories beneath the waves, displacing millions of climate refugees and posing intricate questions regarding national sovereignty when countries lose their physical territory.

The global food systems would undergo a profound transformation. Traditional agricultural regions would experience diminished productivity due to unpredictable shifts in temperature and rainfall patterns. The American Midwest, once regarded as the world's breadbasket, would endure prolonged droughts interspersed with catastrophic floods. Similar challenges would affect other major food-producing regions, including Ukraine's wheat fields and Southeast Asia's rice paddies. While some higher-latitude regions may initially benefit from extended growing seasons, overall global food production would become more volatile and less reliable.

Biodiversity loss would accelerate to unprecedented levels. The Great Barrier Reef, along with most tropical coral reefs, would largely disappear due to ocean acidification and warming, rendering their environments uninhabitable. The Amazon rainforest would approach and potentially surpass a tipping point, transitioning into a savanna and releasing substantial amounts of stored carbon. Numerous species would vanish before our discovery, potentially leading to the loss of medical breakthroughs and crucial links within intricate ecosystems.

During the summer months, the Arctic would undergo a significant transformation, becoming virtually ice-free. This ice-free state would trigger further warming due to the absorption of more heat by dark ocean water compared to reflective ice. Consequently, the thawing of permafrost would accelerate, releasing ancient methane deposits and potentially dormant pathogens. The altered Arctic would profoundly reshape global weather patterns, impacting various regions, including India, where monsoon rains would intensify, and Europe, where winter storms would become more frequent.

Human health would be severely compromised by these changes. Tropical diseases would expand their geographical boundaries into previously temperate areas. Heatwaves would become more frequent and intense, particularly affecting urban regions where the heat island effect exacerbates temperatures. Air quality would deteriorate in many regions due to increased ground-level ozone formation and the proliferation of wildfires. The elderly, young children, and individuals with pre-existing health conditions would be particularly vulnerable to these adverse effects.

Water scarcity would emerge as a critical issue in numerous regions. Major rivers, such as the Ganges and the Yangtze, which are fed by glacial melt, would experience reduced water flow during crucial agricultural seasons. Groundwater depletion would accelerate as drought-stricken regions increasingly rely on aquifers. Water conflicts between nations would intensify, potentially leading to regional instability and forced migration.

Economic systems would face unprecedented challenges. Insurance markets would struggle to cope with the escalating frequency and severity of natural disasters, potentially rendering certain regions uninsurable. Coastal real estate markets would be destabilized by the increasing risk of flooding, leading to significant price fluctuations and potential shortages of essential goods.

The social fabric of numerous societies would be severely tested as climate-related pressures exacerbate existing inequalities. Vulnerable populations, often residing in areas most susceptible to climate impacts and lacking the resources to adapt, would bear a disproportionate burden. Migration patterns would undergo significant shifts as individuals seek refuge from areas becoming uninhabitable due to extreme weather events, potentially straining resources and social services in receiving regions.

The compounding nature of these changes would create feedback loops that could accelerate the pace of change. For instance, the thawing of Arctic permafrost releases greenhouse gases that further warm the climate, leading to more thawing. The loss of reflective ice surfaces results in increased heat absorption, causing further ice loss. Forest fires release carbon dioxide while destroying trees that would otherwise absorb it, creating a vicious cycle.

However, this scenario is not predetermined. The severity of these impacts is directly proportional to the choices we make in the coming years. Every reduction in warming caused by reduced emissions and improved environmental stewardship represents millions of lives improved or saved, species preserved, and ecosystems protected. The technology and knowledge necessary to address these challenges are readily available; what remains is the collective will to implement solutions at the necessary scale and speed.

The primary concern before us is not whether transformations will transpire—they are already underway—but rather the extent to which they will unfold and the extent to which we can effectively adapt to them. Our current actions will determine whether future generations inherit a planet with manageable challenges or catastrophic disruptions. The opportunity for intervention is diminishing, but it has not yet reached a critical juncture.

Helen and James

Helen and James sat in the climate research facility's conference room, still thinking about the big changes they see in 2050. Holographic screens showed the Earth's important signs, showing how much it had changed since they were there in 1900.

"Unbelievable," Helen finally said, her voice deep. "We thought industrial progress would change our world, but we never imagined it would change the whole planet." She looked out the window at the green zone of the facility, next to the dry land outside.

James nodded, adjusting his environmental suit for their outdoor adventures. "The signs were there, weren't they? The smog in London, the industrial waste in our rivers… We just couldn't see the bigger picture." He pulled up their observations on the holographic display, the result of their unique time-traveling perspective.

"The world needs to understand," Helen said, standing up from her chair. "We've been given this incredible chance to see the world over 150 years. The changes we've seen aren't just numbers on a chart; they're the complete transformation of everything we once knew." Her voice trembled as she remembered their visit to her childhood home in Cornwall, now submerged due to rising sea levels.

James approached the window, placing his hand against the temperature-controlled glass. "Recall our expedition to the Arctic last month? Where we once guided expeditions across solid ice, we encountered only open water. The captain of our research vessel informed us that they have not observed persistent ice in those regions for several years."

"And the Amazon," Helen said, pulling up the footage from their South American adventure. "Remember that rainforest we explored when we were last time there? Now, it's just scattered patches among the savannas. All those amazing species we listed, James—they're gone before the world even knew they existed."

They'd been traveling the world for the past year since they woke up, comparing their memories from 1900 with the reality of 2050. Their special perspective had made them super valuable advisors to the Global Climate Response Initiative, giving them insights that no one else could.

"It is very important, we must convey to them that this transformation is not merely about change, but about expediency," James emphasized, his scientific perspective still grappling with the accelerated pace of change. "The Earth has undergone transformations throughout its history, but never at such a rapid clip. The events we have witnessed within a few decades should have transpired over millennia."

Helen commenced compiling their final report, her fingers gracefully traversing the holographic interface. "We must communicate everything—not merely the data, but the narrative. The narrative of how the world we once knew metamorphosed into this one. Perhaps then they will comprehend the significance of the situation."

"And what remains to be salvaged," James added softly. "There is still hope—we have witnessed it as well. The floating gardens of Singapore, the desert reclamation projects in the Sahel, and the artificial reefs safeguarding what remains of our coastlines. Humanity has not relinquished its efforts."

Helen regarded her esteemed friend, perceiving the same resolute spirit in his eyes that had propelled their scientific expeditions decades ago. "Then let us convey this message. Let us make our final contribution to this era be a warning and an impetus to action. After all, who better to articulate the transformation of Earth than two individuals who have traversed it across centuries?"

In unison, they commenced composing their testimony, bridging the gap between the world they had known and the one they now inhabited. They envisioned that their unique perspective might illuminate the path forward for a planet at the precipice of its destiny.

Index